Theory of Vortex Sound

Theory of Vortex Sound is an introduction to the theory of sound generated by hydrodynamic flows. Starting with a review of elementary theoretical acoustics, the book proceeds to a unified treatment of low Mach number vortex-surface interaction noise in terms of the compact Green's function. Problems are provided at the end of each chapter, many of which can be used for extended student projects, and a whole chapter is devoted to worked examples.

It is designed for a one-semester introductory course at the advanced undergraduate or graduate levels. Great care is taken to explain underlying fluid mechanical and acoustic concepts, and to describe as fully as possible the steps in a complicated derivation.

M.S. Howe has been Professor in the Department of Aerospace and Mechanical Engineering at Boston University since 1992. He is a Fellow of the Institute of Acoustics (U.K.) and of the Acoustical Society of America.

Cambridge Texts in Applied Mathematics

Theory of Vortex Sound

M. S. HOWE

Boston University

CAMBRIDGE
UNIVERSITY PRESS

PUBLISHED BY THE PRESS SYNDICATE OF THE UNIVERSITY OF CAMBRIDGE
The Pitt Building, Trumpington Street, Cambridge, United Kingdom

CAMBRIDGE UNIVERSITY PRESS
The Edinburgh Building, Cambridge CB2 2RU, UK
40 West 20th Street, New York, NY 10011-4211, USA
477 Williamstown Road, Port Melbourne, VIC 3207, Australia
Ruiz de Alarcón 13, 28014 Madrid, Spain
Dock House, The Waterfront, Cape Town 8001, South Africa

http://www.cambridge.org

First published 2003

Printed in the United Kingdom at the University Press, Cambridge

Typeface Times Roman 10/13 pt.   *System* LaTeX   [TB]

A catalog record for this book is available from the British Library.

Library of Congress Cataloging in Publication Data
Howe, M. S.
Theory of vortex sound / M. S. Howe.
p.   cm. – (Cambridge texts in applied mathematics)
Includes bibliographical references and index.
ISBN 0-521-81281-X – ISBN 0-521-01223-6 (pbk.)
1. Fluid dynamics.   2. Fluids – Acoustic properties.   I. Title.   II. Series.
TA357.H69   2003
620.1′064 – dc21   2002022280

ISBN 0 521 81281 X hardback
ISBN 0 521 01223 6 paperback

To Shôn Ffowcs Williams

Contents

Preface

Vortex sound is the branch of fluid mechanics concerned with the conversion of hydrodynamic (*rotational*) kinetic energy into the longitudinal disturbances we call sound. The subject is itself a subsection of the theory of aerodynamic sound, which encompasses a much wider range of problems also involving, for example, combustion and 'entropy' sources of sound. The book is based on an introductory one-semester graduate level course given on several occasions at Boston University. Most students at this level possess an insufficient grasp of basic principles to appreciate the subtle coupling of the hydrodynamic and acoustic fields, and many are ill-equipped to deal with the novel analytical techniques that have been developed to investigate the coupling. Great care has therefore been taken to discuss underlying fluid mechanical and acoustic concepts, and to explain as fully as possible the steps in a complicated derivation.

A considerable number of practical problems occur at low Mach numbers (say, less than about 0.4). It seems reasonable, therefore, to confine an introductory discussion specifically to low Mach number flows. It is then possible to investigate a number of idealized hydrodynamic flows involving elementary distributions of vorticity adjacent to solid boundaries, and to analyze in detail the sound produced by these vortex–surface interactions. For a broad range of such problems, and a corresponding broad range of noise problems encountered in industrial applications, the effective acoustic sources turn out to be localized to one or more regions that are small compared to the acoustic wavelength. This permits the development of a unified theory of sound production by vortex–surface interactions in terms of the compact Green's function, culminating in a routine procedure for estimating the sound, and providing, at the same time, an easy identification of those parts of a structure that are likely to be important sources of sound. Many examples of this type are discussed, and they are simple enough for the student to acquire an intuitive understanding of the method of solution and the underlying physics. By these means the reader is encouraged to investigate both the hydrodynamics and the sound generated by a simple flow. Experience has shown that the successful completion of this kind of project, involving the implementation of a widely applicable yet standard procedure for the prediction of sound generation at low Mach numbers, motivates a student to understand the ostensibly difficult parts of the theory. One or more of the problems appended to some of the later chapters can form the basis of a project. The final chapter contains a set of worked examples that have been investigated by students at Boston University. I wish to thank my former students H. Abou-Hussein, A. DeBenedictis, N. Harrison, M. Kim, M. A. Rodrigues, and F. Zagadou for their considerable help in preparing that chapter.

The mathematical ability assumed of the reader is roughly equivalent to that taught in an advanced undergraduate course on Engineering Mathematics. In particular, the reader should be familiar with basic vector differential and integral calculus and with the repeated suffix summation convention of Cartesian tensors (but a detailed knowledge of tensor calculus is not required). An elementary understanding of the properties of the Dirac δ function is desirable (Lighthill, 1958), including its interpretation as the formal limit of an ϵ-sequence, such as

$$\delta(x) = \frac{\epsilon}{\pi(x^2 + \epsilon^2)}, \quad \epsilon \to +0.$$

Much use is made of the formula

$$\delta(f(x)) = \sum_n \frac{\delta(x - x_n)}{|f'(x_n)|},$$

where the summation is over real simple roots of $f(x) = 0$.

M. S. Howe

1

Introduction

1.1 What is Vortex Sound?

Vortex sound is the sound produced as a by-product of unsteady fluid motions (Fig. 1.1.1). It is part of the more general subject of aerodynamic sound. The modern theory of aerodynamic sound was pioneered by James Lighthill in the early 1950s. Lighthill (1952) wanted to understand the mechanisms of noise generation by the jet engines of new passenger jet aircraft that were then about to enter service. However, it is now widely recognized that *any* mechanism that produces sound can actually be formulated as a problem of aerodynamic sound. Thus, apart from the high speed turbulent jet – which may be regarded as a distribution of intense turbulence velocity fluctuations that generate sound by converting a tiny fraction of the jet *rotational* kinetic energy into the longitudinal waves that constitute sound – colliding solid bodies, aeroengine rotor blades, vibrating surfaces, complex fluid–structure interactions in the larynx (responsible for speech), musical instruments, conventional loudspeakers, crackling paper, explosions, combustion and combustion instabilities in rockets, and so forth all fall within the theory of aerodynamic sound in its broadest sense.

In this book we shall consider principally the production of sound by unsteady motions of a fluid. Any fluid that possesses intrinsic kinetic energy, that is, energy not directly attributable to a moving boundary (which is largely withdrawn from the fluid when the boundary motion ceases), must possess *vorticity*. We shall see that in a certain sense and for a vast number of flows vorticity may be regarded as the ultimate source of the sound generated by the flow. Our objective, therefore, is to simplify the general aerodynamic sound problem to obtain a thorough understanding of how this happens, and of how the sound can be estimated quantitatively.

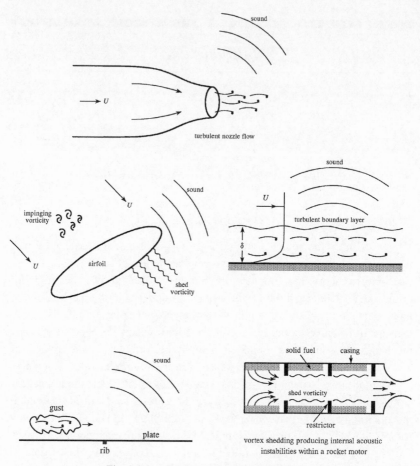

Fig. 1.1.1.   Typical vortex sound problems.

1.2 Equations of Motion of a Fluid

At time t and position $\mathbf{x} = (x_1, x_2, x_3)$, the state of a fluid is defined when the velocity $\mathbf{v}$ and any two thermodynamic variables are specified. Five scalar equations are therefore required to determine the motion. These equations are statements of the conservation of mass, momentum, and energy.

1.2.1 Equation of Continuity

Conservation of mass requires the rate of increase of the fluid mass within a fixed region of space V to be equal to the net influx due to convection across the boundaries of V. The velocity $\mathbf{v}$ and the fluid density ρ must therefore satisfy the equation of continuity, which has the following equivalent forms

$$
\left.\begin{aligned}
\frac{\partial \rho}{\partial t} + \text{div}(\rho \mathbf{v}) &= 0, \\
\frac{1}{\rho} \frac{D\rho}{Dt} + \text{div } \mathbf{v} &= 0, \\
\text{div } \mathbf{v} &= \rho \frac{D}{Dt}\left(\frac{1}{\rho}\right)
\end{aligned}\right\}, \tag{1.2.1}
$$

where

$$
\frac{D}{Dt} = \frac{\partial}{\partial t} + \mathbf{v}\cdot\nabla \equiv \frac{\partial}{\partial t} + v_j \frac{\partial}{\partial x_j} \tag{1.2.2}
$$

is the *material derivative;* the repeated suffix j implies summation over $j = 1, 2, 3$. The last of Equations (1.2.1) states that div $\mathbf{v}$ is equal to the rate of change of fluid volume per unit volume following the motion of the fluid. For an incompressible fluid this is zero, i.e., div $\mathbf{v} = 0$.

1.2.2 Momentum Equation

The momentum equation is also called the *Navier–Stokes equation;* it expresses the rate of change of momentum of a fluid particle in terms of the pressure p, the **viscous** or frictional force, and body forces $\mathbf{F}$ per unit volume. We consider only *Stokesian fluids* (most liquids and monatomic gases, but also a good approximation in air for calculating the frictional drag at a solid boundary) for which the principal frictional forces are expressed in terms of the shear coefficient of viscosity η, which we shall invariably assume to be constant. Then the momentum equation is

$$
\rho \frac{D\mathbf{v}}{Dt} = -\nabla p + \eta \left(\nabla^2 \mathbf{v} + \frac{1}{3}\nabla(\text{div } \mathbf{v})\right) + \mathbf{F}. \tag{1.2.3}
$$

Values of ρ, η and $\nu = \eta/\rho$ (the 'kinematic' viscosity) for air and water at $10\,°\text{C}$ and one atmosphere pressure are given in the Table 1.2.1:

Table 1.2.1. *Density and viscosity*

	ρ, kg/m^3	η, kg/ms	ν, m^2/s
Air	1.23	1.764×10^{-5}	1.433×10^{-5}
Water	1000	1.284×10^{-3}	1.284×10^{-6}

1.2.3 Energy Equation

This equation must be used in its full generality in problems where energy is transferred by heat conduction, where frictional dissipation of sound is occurring, when shock waves are formed by highly nonlinear events, or when sound is being generated by combustion and other heat sources. For our purposes it will usually be sufficient to suppose the flow to be *homentropic;* namely, the specific **entropy** s of the fluid is uniform and constant throughout the fluid, so that the energy equation becomes

$$s = \text{constant.} \tag{1.2.4}$$

We may then assume that the pressure and density are related by an equation of the form

$$p = p(\rho, s), \quad s = \text{constant.} \tag{1.2.5}$$

This equation will be satisfied by both the mean (undisturbed) and unsteady components of the flow. Thus, for an ideal gas

$$p = \text{constant} \times \rho^{\gamma}, \quad \gamma = \text{ratio of specific heats.} \tag{1.2.6}$$

1.3 Equation of Linear Acoustics

The intensity of a sound pressure p in air is usually measured on a decibel scale by the quantity

$$20 \times \log_{10}\left(\frac{|p|}{p_{\text{ref}}}\right),$$

where the reference pressure $p_{\text{ref}} = 2 \times 10^{-5}$ N/m^2. Thus, $p = p_0 \equiv 1$ atmosphere ($= 10^5$ N/m^2) is equivalent to 194 dB. A very loud sound $\sim$120 dB corresponds to

$$\frac{p}{p_0} \approx \frac{2 \times 10^{-5}}{10^5} \times 10^{\left(\frac{120}{20}\right)} = 2 \times 10^{-4} \ll 1.$$

Similarly, for a 'deafening' sound of 160 dB, $p/p_0 \sim 0.02$. This corresponds to a pressure of about 0.3 lbs/in^2 and is loud enough for nonlinear effects to begin to be important.

The passage of a sound wave in the form of a pressure fluctuation is, of course, accompanied by a back-and-forth motion of the fluid at the *acoustic particle velocity* v, say. We shall see later that

$$\text{acoustic particle velocity} \approx \frac{\text{acoustic pressure}}{\text{mean density} \times \text{speed of sound}}.$$

In air the speed of sound is about 340 m/sec. Thus, at 120 dB $v \sim 5$ cm/sec; at 160 dB $v \sim 5$ m/sec.

In most applications the acoustic amplitude is very small relative to the mean pressure p_0, and sound propagation may be studied by linearizing the equations. To do this we shall first consider sound propagating in a *stationary* inviscid fluid of mean pressure p_0 and density ρ_0; let the departures of the pressure and density from these mean values be denoted by p', ρ', where $p'/p_0 \ll 1$, $\rho'/\rho_0 \ll 1$. The linearized momentum equation (1.2.3) becomes

$$\rho_0 \frac{\partial \mathbf{v}}{\partial t} + \nabla p' = \mathbf{F}. \tag{1.3.1}$$

Before linearizing the continuity equation (1.2.1), we introduce an artificial generalization by inserting a **volume source** distribution $q(\mathbf{x}, t)$ on the right-hand side

$$\frac{1}{\rho} \frac{D\rho}{Dt} + \operatorname{div} \mathbf{v} = q, \tag{1.3.2}$$

where q is the rate of increase of fluid volume per unit volume of the fluid, and might represent, for example, the effect of volume pulsations of a small body in the fluid. The linearized equation is then

$$\frac{1}{\rho_0} \frac{\partial \rho'}{\partial t} + \operatorname{div} \mathbf{v} = q. \tag{1.3.3}$$

Now eliminate $\mathbf{v}$ between (1.3.1) and (1.3.3):

$$\frac{\partial^2 \rho'}{\partial t^2} - \nabla^2 p' = \rho_0 \frac{\partial q}{\partial t} - \operatorname{div} \mathbf{F}. \tag{1.3.4}$$

An equation determining the pressure p' alone in terms of q and $\mathbf{F}$ is obtained by invoking the homentropic relation (1.2.5). In the undisturbed and disturbed states we have

$$p_0 = p(\rho_0, s), \qquad p_0 + p' = p(\rho_0 + \rho', s) \approx p(\rho_0, s) + \left(\frac{\partial p}{\partial \rho}(\rho, s) \right)_0 \rho',$$

$$s = \text{constant.} \tag{1.3.5}$$

The derivative is evaluated at the undisturbed values of the pressure and density (p_0, ρ_0). It has the dimensions of velocity^2, and its square root defines the *speed of sound*

$$c_0 = \sqrt{\left(\frac{\partial p}{\partial \rho}\right)_s},$$ (1.3.6)

where the derivative is taken with the entropy s held fixed at its value in the undisturbed fluid. The implication is that losses due to heat transfer between neighboring fluid particles by viscous and thermal diffusion are neglected during the passage of a sound wave (i.e., that the motion of a fluid particle is *adiabatic*).

From (1.3.5): $\rho' = p'/c_0^2$. Hence, substituting for ρ' in (1.3.4), we obtain

$$\left(\frac{1}{c_0^2}\frac{\partial^2}{\partial t^2} - \nabla^2\right)p = \rho_0\frac{\partial q}{\partial t} - \operatorname{div}\mathbf{F},$$ (1.3.7)

where the prime ($'$) on the acoustic pressure has been discarded. This equation governs the production of sound waves by the volume source q and the force $\mathbf{F}$. When these terms are absent the equation describes sound propagation from sources on the boundaries of the fluid, such as the vibrating cone of a loudspeaker.

The volume source q and the body force $\mathbf{F}$ would never appear in a complete description of sound generation within a fluid. They are introduced only when we *think* we understand how to model the real sources of sound in terms of volume sources and forces. In general this can be a dangerous procedure because, as we shall see, small errors in specifying the sources of sound in a fluid can lead to very large errors in the predicted sound. This is because only a tiny fraction of the available energy of a vibrating fluid or structure actually radiates away as sound.

When $\mathbf{F} = \mathbf{0}$, Equation (1.3.1) implies the existence of a velocity potential φ such that $\mathbf{v} = \nabla\varphi$, in terms of which the perturbation pressure is given by

$$p = -\rho_0\frac{\partial\varphi}{\partial t}.$$ (1.3.8)

It follows from this and (1.3.7) (with $\mathbf{F} = \mathbf{0}$) that the velocity potential is the solution of

$$\left(\frac{1}{c_0^2}\frac{\partial^2}{\partial t^2} - \nabla^2\right)\varphi = -q(\mathbf{x}, t).$$ (1.3.9)

This is the wave equation of classical acoustics.

Table 1.3.1. *Speed of sound and acoustic wavelength*

	c_0				λ at 1 kHz	
	m/s	ft/s	km/h	mi/h	m	ft
Air	340	1100	1225	750	0.3	1
Water	1500	5000	5400	3400	1.5	5

For future reference, Table 1.3.1 lists the approximate speeds of sound in air and in water, and the corresponding acoustic wavelength λ at a frequency of 1 kHz (sound of frequency f has wavelength $\lambda = c_0/f$).

1.4 The Special Case of an Incompressible Fluid

Small (adiabatic) pressure and density perturbations δp and $\delta \rho$ satisfy

$$\frac{\delta p}{\delta \rho} \approx c_0^2.$$

In an incompressible fluid the pressure can change by the action of external forces (moving boundaries, etc.), but the density must remain fixed. Thus, $c_0 = \infty$, and Equation (1.3.9) reduces to

$$\nabla^2 \varphi = q(\mathbf{x}, t). \tag{1.4.1}$$

1.4.1 Pulsating Sphere

Consider the motion produced by small amplitude radial pulsations of a sphere of mean radius a. Let the center of the sphere be at the origin, and let its normal velocity be $v_n(t)$. There are no sources within the fluid, so that $q \equiv 0$. Therefore,

$$\left. \begin{array}{ll} \nabla^2 \varphi = 0, & r > a, \\ \partial \varphi / \partial r = v_n(t), & r = a \end{array} \right\} \quad \text{where } r = |\mathbf{x}|.$$

The motion is obviously radially symmetric, so that

$$\nabla^2 \varphi = \frac{1}{r^2} \frac{\partial}{\partial r} \left(r^2 \frac{\partial}{\partial r} \right) \varphi = 0, \quad r > a.$$

Hence,

$$\varphi = \frac{A}{r} + B,$$

where $A \equiv A(t)$ and $B \equiv B(t)$ are functions of t. $B(t)$ can be discarded because the pressure fluctuations ($\sim -\rho_0 \partial\varphi/\partial t$) must vanish as $r \to \infty$. Applying the condition $\partial\varphi/\partial r = v_n$ at $r = a$, we then find

$$\varphi = -\frac{a^2 v_n(t)}{r}, \quad r > a. \tag{1.4.2}$$

Thus, the pressure

$$p = -\rho_0 \frac{\partial\varphi}{\partial t} = \rho_0 \frac{a^2}{r} \frac{dv_n}{dt}(t)$$

decays as $1/r$ with distance from the sphere, and exhibits the unphysical characteristic of changing instantaneously everywhere when dv_n/dt changes its value. For any time t, the volume flux $q(t)$ of fluid is the same across any closed surface enclosing the sphere. Evaluating it for any sphere S of radius $r > a$, as shown in Fig. 1.4.1, we find

$$q(t) = \oint_S \nabla\varphi \cdot d\mathbf{S} = 4\pi a^2 v_n(t),$$

and we may also write

$$\varphi = \frac{-q(t)}{4\pi r}, \quad r > a. \tag{1.4.3}$$

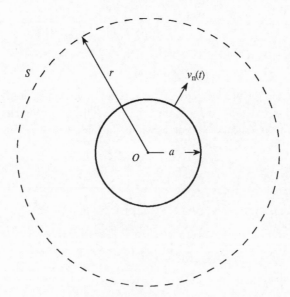

Fig. 1.4.1.

1.4.2 Point Source

The incompressible motion generated by a *volume* point source of strength $q(t)$ at the origin is the solution of

$$\nabla^2 \varphi = q(t)\delta(\mathbf{x}), \quad \text{where } \delta(\mathbf{x}) = \delta(x_1)\delta(x_2)\delta(x_3). \tag{1.4.4}$$

The solution must be radially symmetric and given by

$$\varphi = \frac{A}{r} \quad \text{for } r > 0. \tag{1.4.5}$$

To find A, we integrate (1.4.4) over the interior of a sphere of radius $r = R > 0$, and use the divergence theorem $\int_{r<R} \nabla^2 \varphi \, d^3\mathbf{x} = \oint_S \nabla\varphi \cdot d\mathbf{S}$, where S is the surface of the sphere. Then

$$\oint_S \nabla\varphi \cdot d\mathbf{S} \equiv \left(\frac{-A}{R^2}\right) \times (4\pi R^2) = q(t).$$

Hence, $A = -q(t)/4\pi$ and $\varphi = -q(t)/4\pi r$, which agrees with the solution (1.4.3) for the sphere with the same volume outflow in the region $r > a = $ radius of the sphere. This indicates that when we are interested in modelling the effect of a pulsating sphere at large distances $r \gg a$, it is permissible to replace the sphere by a point source (a monopole) of the same strength $q(t) = $ rate of change of the volume of the sphere. This conclusion is valid for any pulsating body, not just a sphere. However, it is not necessarily a good model (especially when we come to examine the production of *sound* by a pulsating body) in the presence of a *mean fluid flow* past the sphere.

The Solution (1.4.5) for the point source is strictly valid only for $r > 0$, where it satisfies $\nabla^2 \varphi = 0$. What happens as $r \to 0$, where its value is actually undefined? To answer this question, we write the solution in the form

$$\varphi = \lim_{\epsilon \to 0} \frac{-q(t)}{4\pi(r^2 + \epsilon^2)^{\frac{1}{2}}}, \quad \epsilon > 0, \quad \text{in which case } \nabla^2 \varphi = \lim_{\epsilon \to 0} \frac{3\epsilon^2 q(t)}{4\pi(r^2 + \epsilon^2)^{\frac{5}{2}}}.$$

The last limit is just equal to $q(t)\delta(\mathbf{x})$. Indeed when ϵ is small $3\epsilon^2/4\pi(r^2 + \epsilon^2)^{\frac{5}{2}}$ is also small except close to $r = 0$, where it attains a large maximum $\sim 3/4\pi\epsilon^3$. Therefore, for any smoothly varying test function $f(\mathbf{x})$ and any volume V

enclosing the origin

$$\lim_{\epsilon \to 0} \int_V \frac{3\epsilon^2 f(\mathbf{x}) \, d^3\mathbf{x}}{4\pi (r^2 + \epsilon^2)^{\frac{5}{2}}} = f(\mathbf{0}) \lim_{\epsilon \to 0} \int_{-\infty}^{\infty} \frac{3\epsilon^2 \, d^3\mathbf{x}}{4\pi (r^2 + \epsilon^2)^{\frac{5}{2}}}$$

$$= f(\mathbf{0}) \int_0^{\infty} \frac{3\epsilon^2 r^2 \, dr}{(r^2 + \epsilon^2)^{\frac{5}{2}}} = f(\mathbf{0}),$$

where the value of the last integral is independent of ϵ. This is the defining property of the three-dimensional δ function.

Thus, the correct interpretation of the solution

$$\varphi = \frac{-1}{4\pi r} \quad \text{of} \quad \nabla^2 \varphi = \delta(\mathbf{x}) \tag{1.4.6}$$

for a *unit* point source ($q = 1$) is

$$\frac{-1}{4\pi r} = \lim_{\epsilon \to 0} \frac{-1}{4\pi (r^2 + \epsilon^2)^{\frac{1}{2}}}, \quad r \geq 0, \tag{1.4.7}$$

where

$$\nabla^2 \left(\frac{-1}{4\pi r} \right) = \lim_{\epsilon \to 0} \nabla^2 \left(\frac{-1}{4\pi (r^2 + \epsilon^2)^{\frac{1}{2}}} \right) = \lim_{\epsilon \to 0} \frac{3\epsilon^2}{4\pi (r^2 + \epsilon^2)^{\frac{5}{2}}} = \delta(\mathbf{x}). \tag{1.4.8}$$

1.5 Sound Produced by an Impulsive Point Source

The sound generated by the unit, impulsive point source $\delta(\mathbf{x})\delta(t)$ is the solution of

$$\left(\frac{1}{c_0^2} \frac{\partial^2}{\partial t^2} - \nabla^2 \right) \varphi = \delta(\mathbf{x})\delta(t). \tag{1.5.1}$$

The source exists only for an infinitesimal instant of time at $t = 0$; therefore at earlier times $\varphi(\mathbf{x}, t) = 0$ everywhere.

It is evident that the solution is radially symmetric, and that for $r = |\mathbf{x}| > 0$ we have to solve

$$\frac{1}{c_0^2} \frac{\partial^2 \varphi}{\partial t^2} - \frac{1}{r^2} \frac{\partial}{\partial r} \left(r^2 \frac{\partial}{\partial r} \right) \varphi = 0, \quad r > 0. \tag{1.5.2}$$

The identity

$$\frac{1}{r^2} \frac{\partial}{\partial r} \left(r^2 \frac{\partial}{\partial r} \right) \varphi \equiv \frac{1}{r} \frac{\partial^2}{\partial r^2} (r\varphi) \tag{1.5.3}$$

permits us to write Equation (1.5.2) in the form of the *one-dimensional wave equation* for $r\varphi$:

$$\frac{1}{c_0^2}\frac{\partial^2}{\partial t^2}(r\varphi) - \frac{\partial^2}{\partial r^2}(r\varphi) = 0, \quad r > 0. \tag{1.5.4}$$

This has the general solution $r\varphi = \Phi(t - r/c_0) + \Psi(t + r/c_0)$, where Φ and Ψ are arbitrary functions. Hence, the general solution of (1.5.2) is

$$\varphi = \frac{\Phi\left(t - \frac{r}{c_0}\right)}{r} + \frac{\Psi\left(t + \frac{r}{c_0}\right)}{r}, \quad r > 0. \tag{1.5.5}$$

The first term on the right represents a spherically symmetric disturbance that propagates in the direction of increasing values of r at the speed of sound c_0 as t increases, whereas the second represents an *incoming* wave converging toward $\mathbf{x} = \mathbf{0}$. We must therefore set $\Psi = 0$, since it represents sound waves generated at $r = \infty$ that approach the source rather than sound waves generated by the source and radiating away from the source. This is a *causality* or *radiation* condition, that (in the absence of boundaries) sound produced by a source must radiate away from the source. It is also consistent with the Second Law of Thermodynamics, which requires natural systems to change in the more probable direction. An event in which sound waves converge on a point from all directions at infinity is so unlikely as to be impossible in practice; it would be the acoustic analogue of the far-scattered pieces of a broken cup spontaneously reassembling.

To complete the solution it remains to determine the function Φ. We do this by extending the solution down to the source at $r = 0$ by writing (c.f., (1.4.7))

$$\varphi = \frac{\Phi\left(t - \frac{r}{c_0}\right)}{r} = \lim_{\epsilon \to 0} \frac{\Phi\left(t - \frac{r}{c_0}\right)}{(r^2 + \epsilon^2)^{\frac{1}{2}}}, \quad r \geq 0. \tag{1.5.6}$$

Let us substitute this into Equation (1.5.1) and examine what happens as $\epsilon \to 0$. By direct calculation we find

$$\nabla^2\varphi = \frac{1}{r}\frac{\partial^2}{\partial r^2}(r\varphi) = -\frac{3\epsilon^2\Phi(t - r/c_0)}{(r^2 + \epsilon^2)^{\frac{5}{2}}} - \frac{2\epsilon^2\Phi'(t - r/c_0)}{c_0 r(r^2 + \epsilon^2)^{\frac{3}{2}}} + \frac{\Phi''(t - r/c_0)}{c_0^2(r^2 + \epsilon^2)^{\frac{1}{2}}}$$

$$\frac{1}{c_0^2}\frac{\partial^2\varphi}{\partial t^2} = \frac{\Phi''(t - r/c_0)}{c_0^2(r^2 + \epsilon^2)^{\frac{1}{2}}}$$

Therefore,

$$\frac{1}{c_0^2}\frac{\partial^2 \varphi}{\partial t^2} - \nabla^2 \varphi = \frac{3\epsilon^2 \Phi(t - r/c_0)}{(r^2 + \epsilon^2)^{\frac{5}{2}}} + \frac{2\epsilon^2 \Phi'(t - r/c_0)}{c_0 r(r^2 + \epsilon^2)^{\frac{3}{2}}}$$

$$\rightarrow 4\pi \Phi(t)\delta(\mathbf{x}) + 0 \quad \text{as } \epsilon \rightarrow 0, \qquad (1.5.7)$$

where the δ function in the last line follows from (1.4.8), and the '$+ 0$' is obtained by noting that for any smoothly varying test function $f(\mathbf{x})$ and any volume V enclosing the origin

$$\int_V \frac{2\epsilon^2 f(\mathbf{x}) \, d^3\mathbf{x}}{r(r^2 + \epsilon^2)^{\frac{3}{2}}} \approx f(0) \int_{-\infty}^{\infty} \frac{2\epsilon^2 \, d^3\mathbf{x}}{r(r^2 + \epsilon^2)^{\frac{3}{2}}}$$

$$= f(0) \int_0^{\infty} \frac{8\pi\epsilon^2 r \, dr}{(r^2 + \epsilon^2)^{\frac{3}{2}}} = 8\pi\epsilon f(0) \rightarrow 0 \quad \text{as } \epsilon \rightarrow 0.$$

Hence, comparing (1.5.7) with the inhomogeneous wave equation (1.5.1), we find

$$\Phi(t) = \frac{1}{4\pi}\delta(t),$$

and the Solution (1.5.6) becomes

$$\varphi(\mathbf{x}, t) = \frac{1}{4\pi r}\delta\left(t - \frac{r}{c_0}\right) \equiv \frac{1}{4\pi |\mathbf{x}|}\delta\left(t - \frac{|\mathbf{x}|}{c_0}\right). \qquad (1.5.8)$$

This represents a spherical pulse that is nonzero only on the surface of the sphere $r = c_0 t > 0$, whose radius increases at the speed of sound c_0; it vanishes *everywhere* for $t < 0$, before the impulsive source is triggered.

1.6 Free-Space Green's Function

The free-space Green's function $G(\mathbf{x}, \mathbf{y}, t - \tau)$ is the *causal* solution of the wave equation generated by the impulsive point source $\delta(\mathbf{x} - \mathbf{y})\delta(t - \tau)$, located at the point $\mathbf{x} = \mathbf{y}$ at time $t = \tau$. The formula for G is obtained from the solution (1.5.8) for a source at $\mathbf{x} = 0$ at $t = 0$ simply by replacing $\mathbf{x}$ by $\mathbf{x} - \mathbf{y}$ and t by $t - \tau$. In other words, if

$$\left(\frac{1}{c_0^2}\frac{\partial^2}{\partial t^2} - \nabla^2\right) G = \delta(\mathbf{x} - \mathbf{y})\delta(t - \tau), \quad \text{where } G = 0 \quad \text{for } t < \tau, \quad (1.6.1)$$

then

$$G(\mathbf{x}, \mathbf{y}, t - \tau) = \frac{1}{4\pi |\mathbf{x} - \mathbf{y}|}\delta\left(t - \tau - \frac{|\mathbf{x} - \mathbf{y}|}{c_0}\right). \qquad (1.6.2)$$

This represents an impulsive, spherically symmetric wave expanding from the source at **y** at the speed of sound. The wave amplitude decreases inversely with distance $|\mathbf{x} - \mathbf{y}|$ from the source point **y**.

Green's function is the fundamental building block for forming solutions of the inhomogeneous wave equation (1.3.7) of linear acoustics. Let us write this equation in the form

$$\left(\frac{1}{c_0^2} \frac{\partial^2}{\partial t^2} - \nabla^2 \right) p = \mathcal{F}(\mathbf{x}, t), \tag{1.6.3}$$

where the generalized source $\mathcal{F}(\mathbf{x}, t)$ is assumed to be generating waves that propagate away from the source region, in accordance with the radiation condition.

This source distribution can be regarded as a distribution of impulsive point sources of the type on the right of Equation (1.6.1), because

$$\mathcal{F}(\mathbf{x}, t) = \iint_{-\infty}^{\infty} \mathcal{F}(\mathbf{y}, \tau)\, \delta(\mathbf{x} - \mathbf{y})\, \delta(t - \tau)\, d^3\mathbf{y}\, d\tau.$$

The outgoing wave solution for each constituent source of strength

$$\mathcal{F}(\mathbf{y}, \tau)\delta(\mathbf{x} - \mathbf{y})\delta(t - \tau)\, d^3\mathbf{y}\, d\tau \quad \text{is} \quad \mathcal{F}(\mathbf{y}, \tau)G(\mathbf{x}, \mathbf{y}, t - \tau)\, d^3\mathbf{y}\, d\tau,$$

so that by adding up these individual contributions we obtain

$$p(\mathbf{x}, t) = \iint_{-\infty}^{\infty} \mathcal{F}(\mathbf{y}, \tau)G(\mathbf{x}, \mathbf{y}, t - \tau)\, d^3\mathbf{y}\, d\tau \tag{1.6.4}$$

$$= \frac{1}{4\pi} \iint_{-\infty}^{\infty} \frac{\mathcal{F}(\mathbf{y}, \tau)}{|\mathbf{x} - \mathbf{y}|} \delta\left(t - \tau - \frac{|\mathbf{x} - \mathbf{y}|}{c_0} \right) d^3\mathbf{y}\, d\tau \tag{1.6.5}$$

i.e., $p(\mathbf{x}, t) = \dfrac{1}{4\pi} \displaystyle\int_{-\infty}^{\infty} \dfrac{\mathcal{F}\left(\mathbf{y}, t - \frac{|\mathbf{x}-\mathbf{y}|}{c_0}\right)}{|\mathbf{x} - \mathbf{y}|}\, d^3\mathbf{y}.$ \hfill (1.6.6)

The integral formula (1.6.6) is called a **retarded potential;** it represents the pressure at position **x** and time t as a linear superposition of contributions from sources at positions **y**, which radiated at the earlier times $t - |\mathbf{x}-\mathbf{y}|/c_0$, $|\mathbf{x}-\mathbf{y}|/c_0$ being the time of travel of sound waves from **y** to **x**.

1.7 Monopoles, Dipoles, and Quadrupoles

A *volume* point source $q(t)\delta(\mathbf{x})$ of the type considered in Section 1.4 as a model for a pulsating sphere is also called a *monopole* point source. For a compressible medium the corresponding velocity potential it produces is the solution of the equation

$$\left(\frac{1}{c_0^2}\frac{\partial^2}{\partial t^2} - \nabla^2\right)\varphi = -q(t)\delta(\mathbf{x}). \quad (1.7.1)$$

The solution can be written down by analogy with the Solution (1.6.6) of Equation (1.6.3) for the acoustic pressure. Replace p by φ in (1.6.6) and set $\mathcal{F}(\mathbf{y}, \tau) = -q(\tau)\delta(\mathbf{y})$. Then,

$$\varphi(\mathbf{x}, t) = \frac{-q\left(t - \frac{|\mathbf{x}|}{c_0}\right)}{4\pi|\mathbf{x}|} \equiv \frac{-q\left(t - \frac{r}{c_0}\right)}{4\pi r}. \quad (1.7.2)$$

This differs from the corresponding solution (1.4.3) for a pulsating sphere or volume point source in an *incompressible* fluid by the dependence on the *retarded time* $t - \frac{r}{c_0}$. This is physically more realistic; any effects associated with changes in the motion of the sphere (i.e., in the value of the volume outflow rate $q(t)$) are now communicated to a fluid element at distance r after an appropriate time delay r/c_0 required for sound to travel outward from the source.

1.7.1 The Point Dipole

Let $\mathbf{f} = \mathbf{f}(t)$ be a time-dependent vector. Then a source on the right of the acoustic pressure equation (1.6.3) of the form

$$\mathcal{F}(\mathbf{x}, t) = \operatorname{div}(\mathbf{f}(t)\delta(\mathbf{x})) \equiv \frac{\partial}{\partial x_j}(f_j(t)\delta(\mathbf{x})) \quad (1.7.3)$$

is called a *point dipole* (located at the origin). As explained in the Preface, a repeated italic subscript, such as j in this equation, implies a summation over $j = 1, 2, 3$. Equation (1.3.7) shows that the point dipole is equivalent to a force distribution $\mathbf{F}(t) = -\mathbf{f}(t)\delta(\mathbf{x})$ per unit volume applied to the fluid at the origin.

The sound produced by the dipole can be calculated from (1.6.6), but it is easier to use (1.6.5):

$$p(\mathbf{x}, t) = \frac{1}{4\pi}\iint_{-\infty}^{\infty}\frac{\partial}{\partial y_j}(f_j(\tau)\delta(\mathbf{y}))\frac{\delta\left(t - \tau - \frac{|\mathbf{x}-\mathbf{y}|}{c_0}\right)}{|\mathbf{x}-\mathbf{y}|}\,d^3\mathbf{y}\,d\tau.$$

Integrate by parts with respect to each y_j (recalling that $\delta(\mathbf{y}) = 0$ at $y_j = \pm\infty$), and note that

$$\frac{\partial}{\partial y_j}\frac{\delta\left(t - \tau - \frac{|\mathbf{x}-\mathbf{y}|}{c_0}\right)}{|\mathbf{x}-\mathbf{y}|} = -\frac{\partial}{\partial x_j}\frac{\delta\left(t - \tau - \frac{|\mathbf{x}-\mathbf{y}|}{c_0}\right)}{|\mathbf{x}-\mathbf{y}|}.$$

Then,

$$p(\mathbf{x}, t) = \frac{1}{4\pi} \iint_{-\infty}^{\infty} f_j(\tau)\delta(\mathbf{y}) \frac{\partial}{\partial x_j} \left(\frac{\delta\left(t - \tau - \frac{|\mathbf{x}-\mathbf{y}|}{c_0}\right)}{|\mathbf{x} - \mathbf{y}|} \right) d^3\mathbf{y}\, d\tau$$

$$= \frac{1}{4\pi} \frac{\partial}{\partial x_j} \iint_{-\infty}^{\infty} f_j(\tau)\delta(\mathbf{y}) \left(\frac{\delta\left(t - \tau - \frac{|\mathbf{x}-\mathbf{y}|}{c_0}\right)}{|\mathbf{x} - \mathbf{y}|} \right) d^3\mathbf{y}\, d\tau.$$

Thus,

$$p(\mathbf{x}, t) = \frac{\partial}{\partial x_j} \left(\frac{f_j\left(t - \frac{|\mathbf{x}|}{c_0}\right)}{4\pi |\mathbf{x}|} \right). \tag{1.7.4}$$

The same procedure shows that for a *distributed* dipole source of the type $\mathcal{F}(\mathbf{x}, t) = \operatorname{div} \mathbf{f}(\mathbf{x}, t)$ on the right of Equation (1.6.3), the acoustic pressure becomes

$$p(\mathbf{x}, t) = \frac{1}{4\pi} \frac{\partial}{\partial x_j} \int_{-\infty}^{\infty} \frac{f_j\left(\mathbf{y}, t - \frac{|\mathbf{x}-\mathbf{y}|}{c_0}\right)}{|\mathbf{x} - \mathbf{y}|} d^3\mathbf{y}. \tag{1.7.5}$$

A point dipole at the origin orientated in the direction of a unit vector $\mathbf{n}$ is entirely equivalent to two point monopoles of equal but opposite strengths placed a short distance apart (much smaller than the acoustic wavelength) on opposite sides of the origin on a line through the origin parallel to $\mathbf{n}$. For example, if $\mathbf{n}$ is parallel to the x axis, and the sources are distance ϵ apart, the two monopoles would be

$$q(t)\delta\left(x - \frac{\epsilon}{2}\right)\delta(y)\delta(z) - q(t)\delta\left(x + \frac{\epsilon}{2}\right)\delta(y)\delta(z)$$

$$\approx -\epsilon q(t)\delta'(x)\delta(y)\delta(z) \equiv -\frac{\partial}{\partial x}(\epsilon q(t)\delta(\mathbf{x})).$$

This is a fluid *volume* dipole. The relation $p = -\rho_0 \partial\varphi/\partial t$ implies that the equivalent dipole source in the pressure equation (1.3.7) or (1.6.3) is

$$-\rho_0 \frac{\partial}{\partial x}(\epsilon \dot{q}(t)\delta(\mathbf{x})),$$

where the dot denotes differentiation with respect to time.

1.7.2 Quadrupoles

A source distribution involving two space derivatives is equivalent to a combination of four monopole sources (whose net volume source strength is zero), and is called a *quadrupole*. A general quadrupole is a source of the form

$$\mathcal{F}(\mathbf{x}, t) = \frac{\partial^2 T_{ij}}{\partial x_i \partial x_j}(\mathbf{x}, t) \tag{1.7.6}$$

in Equation (1.6.3). The argument above leading to Expression (1.7.5) can be applied twice to show that the corresponding acoustic pressure is given by

$$p(\mathbf{x}, t) = \frac{1}{4\pi} \frac{\partial^2}{\partial x_i \partial x_j} \int_{-\infty}^{\infty} \frac{T_{ij}(\mathbf{y}, t - |\mathbf{x} - \mathbf{y}|/c_0)}{|\mathbf{x} - \mathbf{y}|} d^3\mathbf{y}. \tag{1.7.7}$$

1.7.3 Vibrating Sphere

Let a rigid sphere of radius a execute small amplitude oscillations at speed $U(t)$ in the x_1 direction (Fig. 1.7.1a). Take the coordinate origin at the mean position of the center. In Section 3.5, we shall prove that the motion induced in an ideal fluid when the sphere is *small* is equivalent to that produced by a point volume dipole of strength $2\pi a^3 U(t)$ at its center directed along the x_1 axis, determined

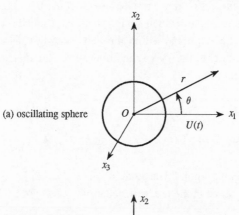

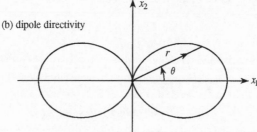

Fig. 1.7.1.

by the solution of

$$\left(\frac{1}{c_0^2}\frac{\partial^2}{\partial t^2} - \nabla^2\right)\varphi = \frac{\partial}{\partial x_1}(2\pi a^3 U(t)\delta(\mathbf{x})). \tag{1.7.8}$$

By analogy with (1.7.3) and (1.7.4), we have

$$\varphi(\mathbf{x}, t) = \frac{\partial}{\partial x_1}\left(\frac{2\pi a^3 U\left(t - \frac{|\mathbf{x}|}{c_0}\right)}{4\pi|\mathbf{x}|}\right). \tag{1.7.9}$$

Now,

$$\frac{\partial}{\partial x_j}|\mathbf{x}| = \frac{x_j}{|\mathbf{x}|}. \tag{1.7.10}$$

Applying this formula for $j = 1$, we find (putting $r = |\mathbf{x}|$ and $x_1 = r\cos\theta$)

$$\varphi = -\underbrace{\frac{a^3\cos\theta}{2r^2}U\left(t - \frac{r}{c_0}\right)}_{\text{near field}} - \underbrace{\frac{a^3\cos\theta}{2c_0 r}\frac{\partial U}{\partial t}\left(t - \frac{r}{c_0}\right)}_{\text{far field}}.$$

The near-field term is dominant at sufficiently small distances r from the origin such that

$$\frac{1}{r} \gg \frac{1}{c_0}\frac{1}{U}\frac{\partial U}{\partial t} \sim \frac{f}{c_0},$$

where f is the characteristic frequency of the oscillations of the sphere. But, sound of frequency f travels a distance

$$c_0/f = \lambda \equiv \text{one acoustic wavelength}$$

in one period of oscillation $1/f$. Hence, the near-field term is dominant when

$$r \ll \lambda.$$

The motion becomes incompressible when $c_0 \to \infty$. In this limit the solution reduces entirely to the near-field term, which is also called the *hydrodynamic near field*; it decreases in amplitude like $1/r^2$ as $r \to \infty$.

The far field is the acoustic region that only exists when the fluid is compressible. It consists of propagating sound waves, carrying energy away from the sphere, and takes over from the near field when $r \gg \lambda$. There is an intermediate zone where $r \sim \lambda$ in which the solution is in a state of transition from the near to the far field. The analytical model (1.7.8), in which the sphere is replaced by a point dipole at its center, involves the implicit assumption that the motion close to the sphere is essentially the same as if the fluid is incompressible. It follows from what we have just said that $a \ll \lambda$, that is, the diameter of the sphere is much smaller than the acoustic wavelength. In general, a body is said to be **acoustically compact** when its characteristic dimension is small compared to the wavelengths of the sound waves it is producing or with which it interacts.

The *intensity* of the sound generated by the sphere in the far field is proportional to φ^2:

$$\varphi^2 \to \frac{a^6}{4c_0^2 r^2} \left(\frac{\partial U}{\partial t} \left(t - \frac{r}{c_0} \right) \right)^2 \cos^2 \theta.$$

The dependence on θ determines the *directivity* of the sound. For the dipole it has the figure of eight pattern illustrated in Fig. 1.7.1b, with peaks in directions parallel to the dipole axis ($\theta = 0, \pi$); there are radiation nulls at $\theta = \frac{\pi}{2}$ (the curve should be imagined to be rotated about the x_1 axis).

1.8 Acoustic Energy Flux

At large distances r from a source region we generally have

$$p(\mathbf{x}, t) \sim \frac{\rho_0 \Phi \left(\theta, \phi, t - \frac{r}{c_0} \right)}{r}, \quad r \to \infty, \tag{1.8.1}$$

where the function Φ depends on the nature of the source distribution, and θ and ϕ are polar angles determining the directivity of the sound. From the radial component of the linearized momentum equation

$$\frac{\partial v_r}{\partial t} = -\frac{1}{\rho_0} \frac{\partial p}{\partial r}$$

$$\equiv \frac{1}{r^2} \Phi \left(\theta, \phi, t - \frac{r}{c_0} \right) + \frac{1}{c_0 r} \frac{\partial \Phi}{\partial t} \left(\theta, \phi, t - \frac{r}{c_0} \right). \tag{1.8.2}$$

The first term in the second line can be neglected when $r \to \infty$, and therefore

$$v_r \sim \frac{1}{c_0 r} \Phi \left(\theta, \phi, t - \frac{r}{c_0} \right) \equiv \frac{p}{\rho_0 c_0}. \tag{1.8.3}$$

By considering the θ and ϕ components of the momentum equation we can show that the corresponding velocity components v_θ, v_ϕ, say, decrease faster than $1/r$ as $r \to \infty$. We therefore conclude from this and (1.8.3) that the acoustic particle velocity is normal to the acoustic wavefronts (the spherical surfaces $r = c_0 t$). In other words, sound consists of longitudinal waves in which the fluid particles oscillate backwards and forwards along the local direction of propagation of the sound.

The **acoustic power** Π radiated by a source distribution can be computed from the formula

$$\Pi = \oint_S p v_r \, dS = \oint_S \frac{p^2}{\rho_0 c_0} \, dS, \tag{1.8.4}$$

where the integration is over the surface S of a large sphere of radius r centered on the source region. Because the surface area $= 4\pi r^2$, we only need to know the pressure and velocity correct to order $1/r$ on S in order to evaluate the integral. Smaller contributions (such as that determined by the first term in the second line of (1.8.2)) decrease too fast as r increases to supply a finite contribution to the integral as $r \to \infty$.

In acoustic problems we are therefore usually satisfied if we can calculate the pressure and velocity in the acoustic far field correct to order $1/r$; this will always enable us to determine the radiated sound power. The formula $v_r = p/\rho_0 c_0$ is applicable at large distances from the sources, where the wavefronts can be regarded as *locally plane,* but it is true identically for plane sound waves. In the latter case, and for spherical waves on the surface of the large sphere of Fig. 1.8.1, the quantity

$$I = p v_r = \frac{p^2}{\rho_0 c_0} \tag{1.8.5}$$

is called the **acoustic intensity.** It is the rate of transmission of acoustic energy per unit area of wavefront.

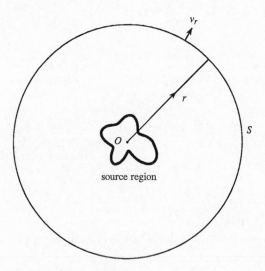

Fig. 1.8.1.

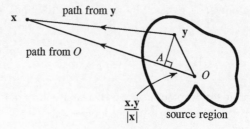

Fig. 1.9.1.

1.9 Calculation of the Acoustic Far Field

We now discuss the approximations necessary to evaluate the sound in the far field from the retarded potential representation:

$$p(\mathbf{x}, t) = \frac{1}{4\pi} \int_{-\infty}^{\infty} \frac{\mathcal{F}\left(\mathbf{y}, t - \frac{|\mathbf{x}-\mathbf{y}|}{c_0}\right)}{|\mathbf{x} - \mathbf{y}|} \, d^3\mathbf{y}. \tag{1.9.1}$$

We assume that $\mathcal{F}(\mathbf{x}, t) \neq 0$ only within a finite source region (Fig. 1.9.1), and take the coordinate origin O within the region.

When $|\mathbf{x}| \to \infty$ and $\mathbf{y}$ lies within the source region (so that $|\mathbf{x}| \gg |\mathbf{y}|$)

$$|\mathbf{x} - \mathbf{y}| \equiv (|\mathbf{x}|^2 - 2\mathbf{x} \cdot \mathbf{y} + |\mathbf{y}|^2)^{\frac{1}{2}} = |\mathbf{x}| \left\{ 1 - \frac{2\mathbf{x} \cdot \mathbf{y}}{|\mathbf{x}|^2} + \frac{|\mathbf{y}|^2}{|\mathbf{x}|^2} \right\}^{\frac{1}{2}}$$

$$\approx |\mathbf{x}| \left\{ 1 - \frac{\mathbf{x} \cdot \mathbf{y}}{|\mathbf{x}|^2} + O\left(\frac{|\mathbf{y}|^2}{|\mathbf{x}|^2}\right) \right\}$$

Then,

$$|\mathbf{x} - \mathbf{y}| \approx |\mathbf{x}| - \frac{\mathbf{x} \cdot \mathbf{y}}{|\mathbf{x}|} \quad \text{when} \quad \frac{|\mathbf{y}|}{|\mathbf{x}|} \ll 1. \tag{1.9.2}$$

Also,

$$\frac{1}{|\mathbf{x} - \mathbf{y}|} \approx \frac{1}{\left(|\mathbf{x}| - \frac{\mathbf{x} \cdot \mathbf{y}}{|\mathbf{x}|}\right)} \approx \frac{1}{|\mathbf{x}|} \left(1 + \frac{\mathbf{x} \cdot \mathbf{y}}{|\mathbf{x}|^2}\right)$$

Therefore, $\dfrac{1}{|\mathbf{x} - \mathbf{y}|} \approx \dfrac{1}{|\mathbf{x}|} + \dfrac{\mathbf{x} \cdot \mathbf{y}}{|\mathbf{x}|^3} \quad \text{when} \quad \dfrac{|\mathbf{y}|}{|\mathbf{x}|} \ll 1. \tag{1.9.3}$

The Approximation (1.9.3) shows that, in order to obtain the far-field approximation of the Solution (1.9.1) that behaves like $1/r = 1/|\mathbf{x}|$ as $|\mathbf{x}| \to \infty$, it is sufficient to replace $|\mathbf{x} - \mathbf{y}|$ in the denominator of the integrand by $|\mathbf{x}|$. However, in the argument of the source strength $\mathcal{F}$ it is important to retain possible *phase differences* between the sound waves generated by components of the source distribution at different locations **y**; we therefore replace $|\mathbf{x} - \mathbf{y}|$ in the retarded time by the right-hand side of (1.9.2). Hence,

$$p(\mathbf{x}, t) \approx \frac{1}{4\pi|\mathbf{x}|} \int_{-\infty}^{\infty} \mathcal{F}\left(\mathbf{y}, t - \frac{|\mathbf{x}|}{c_0} + \frac{\mathbf{x} \cdot \mathbf{y}}{c_0|\mathbf{x}|}\right) d^3\mathbf{y}, \quad |\mathbf{x}| \to \infty. \quad (1.9.4)$$

This is called the *Fraunhofer approximation.*

The source region may extend over many characteristic acoustic wavelengths of the sound. By retaining the contribution $\mathbf{x} \cdot \mathbf{y}/c_0|\mathbf{x}|$ to the retarded time we ensure that any *interference* between waves generated at different positions within the source region is correctly described by the far-field approximation. In Fig. 1.9.1 the acoustic travel time from a source point **y** to the far field point **x** is equal to that from the point labelled A to **x** when $|\mathbf{x}| \to \infty$. The travel time over the distance OA is just $\mathbf{x} \cdot \mathbf{y}/c_0|\mathbf{x}|$, so that $|\mathbf{x}|/c_0 - \mathbf{x} \cdot \mathbf{y}/c_0|\mathbf{x}|$ gives the correct value of the retarded time when $|\mathbf{x}| \to \infty$.

1.9.1 Dipole Source Distributions

By applying the far-field formula (1.9.4) to a dipole source $\mathcal{F}(\mathbf{x}, t) = \text{div } \mathbf{f}(\mathbf{x}, t)$ we obtain (from (1.7.5))

$$p(\mathbf{x}, t) \approx \frac{1}{4\pi} \frac{\partial}{\partial x_j} \left[\frac{1}{|\mathbf{x}|} \int_{-\infty}^{\infty} f_j\left(\mathbf{y}, t - \frac{|\mathbf{x}|}{c_0} + \frac{\mathbf{x} \cdot \mathbf{y}}{c_0|\mathbf{x}|}\right) d^3\mathbf{y} \right]$$

$$\approx \frac{1}{4\pi|\mathbf{x}|} \frac{\partial}{\partial x_j} \int_{-\infty}^{\infty} f_j\left(\mathbf{y}, t - \frac{|\mathbf{x}|}{c_0} + \frac{\mathbf{x} \cdot \mathbf{y}}{c_0|\mathbf{x}|}\right) d^3\mathbf{y}, \quad |\mathbf{x}| \to \infty, \quad (1.9.5)$$

because the differential operator $\partial/\partial x_j$ need not be applied to $1/|\mathbf{x}|$ as this would give a contribution decreasing like $1/r^2$ at large distances from the dipole.

However, it is useful to make a further transformation that replaces $\partial/\partial x_j$ by the time derivative $\partial/\partial t$, which is usually more easily estimated in applications. To do this, we observe that

$$\frac{\partial f_j}{\partial x_j}\left(\mathbf{y}, t - \frac{|\mathbf{x}|}{c_0} + \frac{\mathbf{x} \cdot \mathbf{y}}{c_0|\mathbf{x}|}\right)$$

$$= \frac{\partial f_j}{\partial t}\left(\mathbf{y}, t - \frac{|\mathbf{x}|}{c_0} + \frac{\mathbf{x} \cdot \mathbf{y}}{c_0|\mathbf{x}|}\right) \frac{\partial}{\partial x_j}\left(t - \frac{|\mathbf{x}|}{c_0} + \frac{\mathbf{x} \cdot \mathbf{y}}{c_0|\mathbf{x}|}\right)$$

$$= \frac{\partial f_j}{\partial t}\left(\mathbf{y}, t - \frac{|\mathbf{x}|}{c_0} + \frac{\mathbf{x} \cdot \mathbf{y}}{c_0|\mathbf{x}|}\right) \left(-\frac{x_j}{c_0|\mathbf{x}|} + \frac{y_j}{c_0|\mathbf{x}|} - \frac{(\mathbf{x} \cdot \mathbf{y})x_j}{c_0|\mathbf{x}|^3}\right)$$

$$\approx -\frac{x_j}{c_0|\mathbf{x}|} \frac{\partial f_j}{\partial t}\left(\mathbf{y}, t - \frac{|\mathbf{x}|}{c_0} + \frac{\mathbf{x} \cdot \mathbf{y}}{c_0|\mathbf{x}|}\right) \quad \text{as } |\mathbf{x}| \to \infty.$$

Hence, the far field of a distribution of dipoles $\mathcal{F}(\mathbf{x}, t) = \operatorname{div} \mathbf{f}(\mathbf{x}, t)$ is given by

$$p(\mathbf{x}, t) = \frac{-x_j}{4\pi c_0 |\mathbf{x}|^2} \frac{\partial}{\partial t} \int_{-\infty}^{\infty} f_j \left(\mathbf{y}, t - \frac{|\mathbf{x}|}{c_0} + \frac{\mathbf{x} \cdot \mathbf{y}}{c_0 |\mathbf{x}|} \right) d^3 \mathbf{y}. \qquad (1.9.6)$$

Note that

$$\frac{x_j}{|\mathbf{x}|^2} = \frac{x_j}{|\mathbf{x}|} \frac{1}{|\mathbf{x}|},$$

where $x_j / |\mathbf{x}|$ is the jth component of the unit vector $\mathbf{x}/|\mathbf{x}|$. Thus, the additional factor of $x_j / |\mathbf{x}|$ in (1.9.6) does not change the rate of decay of the sound with distance from the source (which is still like $1/r$), but it does have an influence on the acoustic *directivity*.

A comparison of (1.9.5) and (1.9.6) leads to the following rule for interchanging space and time derivatives *in the acoustic far field:*

$$\frac{\partial}{\partial x_j} \longleftrightarrow -\frac{1}{c_0} \frac{x_j}{|\mathbf{x}|} \frac{\partial}{\partial t}. \qquad (1.9.7)$$

1.9.2 Quadrupole Source Distributions

For the Quadrupole (1.7.6)

$$\mathcal{F}(\mathbf{x}, t) = \frac{\partial^2 T_{ij}}{\partial x_i \partial x_j}(\mathbf{x}, t),$$

and

$$p(\mathbf{x}, t) = \frac{1}{4\pi} \frac{\partial^2}{\partial x_i \partial x_j} \int_{-\infty}^{\infty} \frac{T_{ij}(\mathbf{y}, t - |\mathbf{x} - \mathbf{y}|/c_0)}{|\mathbf{x} - \mathbf{y}|} d^3 \mathbf{y}.$$

By applying (1.9.4) and the rule (1.9.7), we find that the acoustic far field is given by

$$p(\mathbf{x}, t) \approx \frac{x_i x_j}{4\pi c_0^2 |\mathbf{x}|^3} \frac{\partial^2}{\partial t^2} \int_{-\infty}^{\infty} T_{ij} \left(\mathbf{y}, t - \frac{|\mathbf{x}|}{c_0} + \frac{\mathbf{x} \cdot \mathbf{y}}{c_0 |\mathbf{x}|} \right) d^3 \mathbf{y}, \quad |\mathbf{x}| \to \infty. \qquad (1.9.8)$$

1.9.3 Example

For the $(1, 2)$ *point* quadrupole

$$\mathcal{F}(\mathbf{x}, t) = \frac{\partial^2}{\partial x_1 \partial x_2}(T(t)\delta(\mathbf{x}))$$

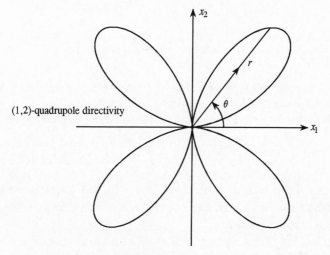

(1,2)-quadrupole directivity

Fig. 1.9.2.

Equation (1.9.8) shows that in the acoustic far field

$$p(\mathbf{x}, t) \approx \frac{x_1 x_2}{4\pi c_0^2 |\mathbf{x}|^3} \frac{\partial^2 T}{\partial t^2}\left(t - \frac{|\mathbf{x}|}{c_0}\right), \quad |\mathbf{x}| \to \infty.$$

If we use spherical polar coordinates, such that

$$x_1 = r\cos\theta, \qquad x_2 = r\sin\theta\cos\phi, \qquad x_3 = r\sin\theta\sin\phi,$$

we can write the pressure in the form

$$p(\mathbf{x}, t) \approx \frac{\sin 2\theta \cos\phi}{8\pi c_0^2 |\mathbf{x}|} \frac{\partial^2 T}{\partial t^2}\left(t - \frac{|\mathbf{x}|}{c_0}\right), \quad |\mathbf{x}| \to \infty.$$

The directivity of the sound ($\propto p^2$) is therefore represented by $\sin^2 2\theta \cos^2\phi$.
Its shape is plotted in Fig. 1.9.2 for radiation in the x_1, x_2 plane ($\phi = 0, \pi$). The
four-lobe cloverleaf pattern is characteristic of a quadrupole T_{ij} for which $i \neq j$.

Problems 1

1. A plane sound wave propagating parallel to the x axis satisfies the equation

$$\left(\frac{1}{c_0^2}\frac{\partial^2}{\partial t^2} - \frac{\partial^2}{\partial x^2}\right)\varphi = 0,$$

with general solution

$$\varphi = \Phi\left(t - \frac{x}{c_0}\right) + \Psi\left(t + \frac{x}{c_0}\right),$$

where Φ and Ψ are arbitrary functions respectively representing waves propagating in the positive and negative x directions.

Show that for a wave propagating in the positive x direction in an ideal gas

$$v = \frac{p}{\rho_0 c_0}, \qquad \rho = \frac{p}{c_0^2}, \qquad T = \frac{p}{\rho_0 c_p},$$

where v is the acoustic particle velocity; p, ρ, and T are respectively the acoustic pressure, density, and temperature variations, and c_p is the specific heat at constant pressure.

2. Calculate the acoustic power (1.8.4) radiated by an acoustically compact sphere of radius R executing small amplitude translational oscillations of frequency ω and velocity $U(t) = U_0 \cos(\omega t)$, where $U_0 = $ constant.

3. As for Problem 2, when the sphere executes small amplitude radial oscillations at normal velocity $v_n = U_0 \cos(\omega t)$, $U_0 = $ constant.

4. A *volume* point source of strength $q_0(t)$ translates at *constant, subsonic* velocity **U**. The velocity potential $\varphi(\mathbf{x}, t)$ of the radiated sound is determined by the solution of

$$\left(\frac{1}{c_0^2}\frac{\partial^2}{\partial t^2} - \nabla^2\right)\varphi = -q_0(t)\delta(\mathbf{x} - \mathbf{U}t).$$

Show that

$$\varphi(\mathbf{x}, t) = \frac{-q_0(t - R/c_0)}{4\pi R(1 - M\cos\Theta)}, \qquad M = \frac{U}{c_0},$$

where R is the distance of the reception point **x** from the source position *at the time of emission* of the sound received at **x** at time t, and Θ is the angle between **U** and the direction of propagation of this sound.

2

Lighthill's Theory

2.1 The Acoustic Analogy

The sound generated by turbulence in an unbounded fluid is usually called *aerodynamic sound*. Most unsteady flows of technological interest are of high Reynolds number and turbulent, and the acoustic radiation is a very small by-product of the motion. The turbulence is usually produced by fluid motion over a solid boundary or by flow instability. Lighthill (1952) transformed the Navier–Stokes and continuity equations to form an exact, inhomogeneous wave equation whose source terms are important only within the turbulent region. He argued that sound is a very small component of the whole motion and that, once generated, its back-reaction on the main flow can usually be ignored. The properties of the unsteady flow in the source region may then be determined by neglecting the production and propagation of the sound, a reasonable approximation if the Mach number M is small, and there are many important flows where the hypothesis is obviously correct, and where the theory leads to unambiguous predictions of the sound.

Lighthill was initially interested in solving the problem, illustrated in Fig. 2.1.1a, of the sound produced by a turbulent nozzle flow. However, his original theory actually applies to the simpler situation shown in Fig. 2.1.1b, in which the sound is imagined to be generated by a finite region of rotational flow in an *unbounded* fluid. This avoids complications caused by the presence of the nozzle. The fluid is assumed to be at rest at infinity, where the mean pressure, density, and sound speed are respectively equal to p_0, ρ_0, c_0. Lighthill compared the equations for the production of acoustic *density* fluctuations in the real flow with those in an ideal linear acoustic medium that coincides with the real fluid at large distances from the sources.

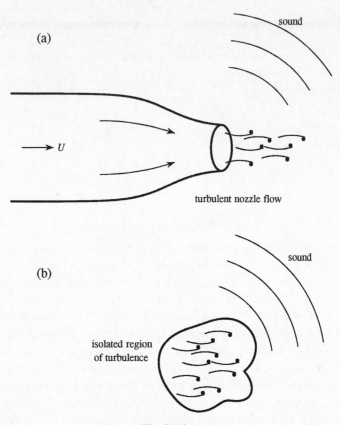

Fig. 2.1.1.

To do this, body forces are neglected, and the ith component of the momentum equation (1.2.3) is cast in the form

$$\rho\frac{\partial v_i}{\partial t} + \rho v_j\frac{\partial v_i}{\partial x_j} = -\frac{\partial p}{\partial x_i} + \frac{\partial \sigma_{ij}}{\partial x_j} \equiv -\frac{\partial}{\partial x_j}(p\delta_{ij} - \sigma_{ij}). \qquad (2.1.1)$$

δ_{ij} is the *Kronecker delta* (= 1 for $i = j$, and 0 for $i \neq j$), and σ_{ij} is the viscous stress tensor defined (for a Stokesian fluid) by

$$\sigma_{ij} = 2\eta\big(e_{ij} - \tfrac{1}{3}e_{kk}\delta_{ij}\big), \qquad (2.1.2)$$

where

$$e_{ij} = \frac{1}{2}\left(\frac{\partial v_i}{\partial x_j} + \frac{\partial v_j}{\partial x_i}\right) \qquad (2.1.3)$$

is the rate of strain tensor. Next multiply the continuity equation (1.2.1) by v_i:

$$v_i \frac{\partial \rho}{\partial t} + v_i \frac{\partial(\rho v_j)}{\partial x_j} = 0.$$

By adding this to Equation (2.1.1), we obtain the *Reynolds form* of the momentum equation

$$\frac{\partial(\rho v_i)}{\partial t} = -\frac{\partial \pi_{ij}}{\partial x_j}, \qquad (2.1.4)$$

where

$$\pi_{ij} = \rho v_i v_j + (p - p_0)\delta_{ij} - \sigma_{ij}, \qquad (2.1.5)$$

is called the **momentum flux** tensor, and the constant pressure p_0 is inserted for convenience.

In an ideal, linear acoustic medium, the momentum flux tensor contains only the pressure

$$\pi_{ij} \rightarrow \pi_{ij}^0 = (p - p_0)\delta_{ij} \equiv c_0^2(\rho - \rho_0)\delta_{ij}, \qquad (2.1.6)$$

and the momentum equation then reduces to

$$\frac{\partial(\rho v_i)}{\partial t} + \frac{\partial}{\partial x_i}\left[c_0^2(\rho - \rho_0)\right] = 0. \qquad (2.1.7)$$

If the continuity equation (1.2.1) is written in the slightly modified form

$$\frac{\partial}{\partial t}(\rho - \rho_0) + \frac{\partial(\rho v_i)}{\partial x_i} = 0, \qquad (2.1.8)$$

we can eliminate the momentum density ρv_i between (2.1.7) and (2.1.8) to obtain the equation of *linear acoustics* satisfied by the perturbation density $\rho - \rho_0$

$$\left(\frac{1}{c_0^2}\frac{\partial^2}{\partial t^2} - \nabla^2\right)\left[c_0^2(\rho - \rho_0)\right] = 0. \qquad (2.1.9)$$

Because the turbulence is neglected in this approximation, and there are no externally applied forces or moving boundaries, the unique solution of this equation that satisfies the radiation condition of *outgoing wave behavior* is simply $\rho - \rho_0 = 0$.

It can now be asserted that the sound generated by the turbulence in the *real fluid* is exactly equivalent to that produced in the ideal, stationary acoustic medium (which is governed by (2.1.9) in turbulence-free regions) forced by the stress distribution

$$
\begin{aligned}
T_{ij} &= \pi_{ij} - \pi_{ij}^0 \\
&= \rho v_i v_j + \left((p - p_0) - c_0^2(\rho - \rho_0)\right)\delta_{ij} - \sigma_{ij},
\end{aligned}
\tag{2.1.10}
$$

where T_{ij} is called the *Lighthill stress tensor*. This is **Lighthill's acoustic analogy.**

Indeed, we can rewrite (2.1.4) as the momentum equation for an ideal, stationary acoustic medium of mean density ρ_0 and sound speed c_0 subject to the externally applied stress T_{ij}

$$
\frac{\partial(\rho v_i)}{\partial t} + \frac{\partial \pi_{ij}^0}{\partial x_j} = -\frac{\partial}{\partial x_j}\left(\pi_{ij} - \pi_{ij}^0\right),
$$

or

$$
\frac{\partial(\rho v_i)}{\partial t} + \frac{\partial}{\partial x_i}\left[c_0^2(\rho - \rho_0)\right] = -\frac{\partial T_{ij}}{\partial x_j}.
\tag{2.1.11}
$$

By eliminating the momentum density ρv_i between this and the continuity equation (2.1.8) (the same procedure used above for the linear problem), we obtain **Lighthill's equation**

$$
\left(\frac{1}{c_0^2}\frac{\partial^2}{\partial t^2} - \nabla^2\right)\left[c_0^2(\rho - \rho_0)\right] = \frac{\partial^2 T_{ij}}{\partial x_i \partial x_j}.
\tag{2.1.12}
$$

This is the *exact,* nonlinear counterpart of (2.1.9). The problem of calculating the turbulence generated sound is therefore equivalent to solving this equation for the radiation into a stationary, ideal fluid produced by a distribution of *quadrupole* sources whose strength per unit volume is the Lighthill stress tensor T_{ij}. The quadrupole character of the turbulence sources is one of the most important conclusions of Lighthill's theory; it implies (see Section 2.2) that free-field turbulence is an extremely weak sound source, and that in a typical low Mach number flow only a tiny fraction of the available flow energy is converted into sound.

In the definition (2.1.10) of T_{ij}, the term $\rho v_i v_j$ is called the *Reynolds stress*. For the simplified problem of Fig. 2.1.1b it is a nonlinear quantity that can be neglected except where the motion is turbulent. The second term represents the *excess* of momentum transfer by the pressure over that in the ideal (linear) fluid of density ρ_0 and sound speed c_0. This is produced by wave amplitude nonlinearity, and by mean density variations in the source flow. The viscous

stress tensor σ_{ij} is linear in the perturbation quantities, and properly accounts
for the attenuation of the sound; in most applications the *Reynolds number*
in the source region is very large, and σ_{ij} can be neglected, and the viscous
attenuation of the radiating sound is usually ignored.

2.2 Lighthill's v^8 Law

The formal solution of Lighthill's equation (2.1.12) with outgoing wave behavior is given by (1.7.7) with $p(\mathbf{x}, t)$ replaced by $c_0^2(\rho - \rho_0)$

$$c_0^2(\rho - \rho_0)(\mathbf{x}, t) = \frac{1}{4\pi} \frac{\partial^2}{\partial x_i \partial x_j} \int_{-\infty}^{\infty} \frac{T_{ij}(\mathbf{y}, t - |\mathbf{x} - \mathbf{y}|/c_0)}{|\mathbf{x} - \mathbf{y}|} d^3\mathbf{y}. \quad (2.2.1)$$

This is strictly an alternative, integral equation representation of Equation
(2.1.12); it provides a useful prediction of the sound only when T_{ij} is known
or has been determined by some other means. This is because the terms in the
definition (2.1.10) of T_{ij} not only account for the generation of sound, but also
govern acoustic *self-modulation* caused by acoustic nonlinearity, the *convection* of sound waves by the turbulent velocity, *refraction* caused by sound speed
variations, and *attenuation* due to thermal and viscous actions. The influence
of acoustic nonlinearity and of thermoviscous dissipation is usually sufficiently
weak to be neglected within the source region, although they may affect propagation to a distant observer. Convection and refraction of sound within and
near the source flow can be important, for example in the presence of a mean
shear layer (when the Reynolds stress will include terms like $\rho U_i u_j$, where $\mathbf{U}$
and $\mathbf{u}$ respectively denote the mean and fluctuating components of $\mathbf{v}$), or when
there are large variations in the mean thermodynamic properties of the medium
within the source region; such effects are described by the presence of unsteady
linear terms in T_{ij} (Ffowcs Williams, 1974).

Thus, to predict the radiated sound from Lighthill's equation (2.2.1) it is
usually necessary to suppose that all of these acoustic effects in the source flow
(which really depend on fluid *compressibility*) are in some sense negligible.
This means that in practice it must be possible to derive a good approximation
for T_{ij} by taking the source flow to be effectively *incompressible*. This is often
possible when the characteristic Mach number $M \sim v/c_0$ is small (specifically,
when $M^2 \ll 1$), and when the wavelength of the sound is much larger than the
size of the source region.

Consider the particular but important case in which the mean density and
sound speed are uniform throughout the fluid. The variations in the density
ρ within a low Mach number, high Reynolds number source flow are then of

order $\rho_0 M^2$ (Batchelor, 1967). Thus, $\rho v_i v_j = \rho_0(1 + O(M^2))v_i v_j \approx \rho_0 v_i v_j$. Similarly, if $c(\mathbf{x}, t)$ is the local speed of sound in the source region, it may also be shown that $c_0^2/c^2 = 1 + O(M^2)$, so that

$$p - p_0 - c_0^2(\rho - \rho_0) \approx (p - p_0)\left(1 - c_0^2/c^2\right) \sim O(\rho_0 v^2 M^2).$$

Hence, if viscous dissipation is neglected we make the approximation

$$T_{ij} \approx \rho_0 v_i v_j, \quad \text{provided that } M^2 \ll 1. \tag{2.2.2}$$

In the acoustic region outside the source flow $c_0^2(\rho - \rho_0) = p - p_0$. If the irrelevant constant pressure p_0 is suppressed, the Solution (2.2.1) of Lighthill's equation therefore becomes

$$\begin{aligned}
p(\mathbf{x}, t) &\approx \frac{\partial^2}{\partial x_i \partial x_j} \int \frac{\rho_0 v_i v_j(\mathbf{y}, t - |\mathbf{x} - \mathbf{y}|/c_0)}{4\pi |\mathbf{x} - \mathbf{y}|} d^3 \mathbf{y} \\
&\approx \frac{x_i x_j}{4\pi c_0^2 |\mathbf{x}|^3} \frac{\partial^2}{\partial t^2} \int \rho_0 v_i v_j \left(\mathbf{y}, t - \frac{|\mathbf{x}|}{c_0} + \frac{\mathbf{x} \cdot \mathbf{y}}{c_0 |\mathbf{x}|}\right) d^3 \mathbf{y}, \quad |\mathbf{x}| \to \infty,
\end{aligned} \tag{2.2.3}$$

where in the second line we have used the formula (1.9.8) for the far field of a quadrupole distribution. Quantitative predictions can be made from this formula provided the behavior of the Reynolds stress $\rho_0 v_i v_j$ is known.

To determine the order of magnitude of p, we introduce a characteristic velocity v and length scale ℓ (of the energy-containing eddies) of the turbulence sources. The value of ℓ depends on the mechanism responsible for turbulence production, such as the width of a jet mixing layer. Fluctuations in $v_i v_j$ occurring in different regions of the turbulent flow separated by distances $> O(\ell)$ will tend to be statistically independent, and the sound may be considered to be generated by a collection of V_0/ℓ^3 independent eddies, where V_0 is the volume occupied by the turbulence (Fig. 2.2.1). The characteristic frequency of the turbulent fluctuations $f \sim v/\ell$, so that the wavelength (c_0/f) of the sound $\sim \ell/M \gg \ell$ (because $M = v/c_0 \ll 1$). Hence, we arrive at the important conclusion that *the turbulence eddies are each acoustically compact*. This means that when the integral in (2.2.3) is confined to a single eddy, the retarded time variations $\mathbf{x} \cdot \mathbf{y}/c_0|\mathbf{x}|$ across that eddy can be neglected; that is, if the coordinate origin is temporarily placed at O within the eddy, we can set $\mathbf{x} \cdot \mathbf{y}/c_0|\mathbf{x}| = 0$ in the integration over that eddy. The value of the integral over the eddy then may be estimated to be of order $\rho_0 v^2 \ell^3$.

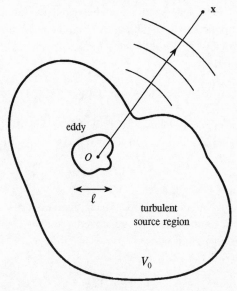

Fig. 2.2.1.

The order of magnitude of the time derivative for changes in the source region is

$$\frac{\partial}{\partial t} \sim \frac{v}{\ell}.$$

Therefore, it follows from (2.2.3) that, for one eddy, the far-field acoustic pressure satisfies

$$p \sim \frac{\ell}{|\mathbf{x}|} \frac{\rho_0 v^4}{c_0^2} = \frac{\ell}{|\mathbf{x}|} \rho_0 v^2 M^2. \tag{2.2.4}$$

The acoustic *power* radiated by the eddy is determined by the surface integral (1.8.4) taken over a large sphere centered on the eddy. Thus, in order of magnitude,

$$\text{acoustic power radiated by one eddy} \sim 4\pi |\mathbf{x}|^2 \frac{p^2}{\rho_0 c_0} \sim \frac{\ell^2 \rho_0 v^8}{c_0^5} = \ell^2 \rho_0 v^3 M^5.$$
$$\tag{2.2.5}$$

This is Lighthill's 'eighth power' law.

The *total* power radiated from the whole of the turbulent region of volume V_0, containing V_0/ℓ^3 independent eddies, is

$$\Pi_q \approx \frac{V_0}{\ell^3} \times (\ell^2 \rho_0 v^3 M^5) = \frac{v}{\ell} \rho_0 v^2 M^5 V_0.$$

Dimensional arguments and experiment indicate that the rate Π_0, say, at which energy must be supplied by the action of external forces to maintain the kinetic energy of a statistically steady turbulent field occupying a volume V_0 is given in order of magnitude by

$$\Pi_0 \sim \frac{v}{\ell} \rho_0 v^2 V_0.$$

Therefore, the mechanical *efficiency* with which turbulence kinetic energy is converted into sound is

$$\frac{\Pi_q}{\Pi_0} \sim M^5. \tag{2.2.6}$$

This is smaller than about 0.01 for Mach numbers $M < 0.4$, confirming Lighthill's hypothesis that the flow generated sound is an infinitesimal by-product of the turbulent motion.

2.3 Curle's Theory

In most applications of Lighthill's theory it is necessary to generalize the solution (2.2.1) to account for the presence of solid bodies in the flow. Indeed, turbulence is frequently generated in the boundary layers and wakes of flow past such bodies (airfoils, flow control surfaces, etc.), and the unsteady surface forces (dipoles) that arise are likely to make a significant contribution to the production of sound. The procedure in such cases is to introduce a system of mathematical **control surfaces** that can be deformed to coincide with the surfaces of the different moving or stationary bodies, although for the moment we shall discuss only cases involving *stationary* bodies. Before doing this we establish an integral transformation formula that is used repeatedly in problems of this kind.

2.3.1 Volume and Surface Integrals

Let V be the fluid *outside* a closed control surface S (Fig. 2.3.1) defined by the equation

$$f(\mathbf{x}) = 0, \quad \text{where} \begin{cases} f(\mathbf{x}) > 0 & \text{for } \mathbf{x} \text{ in } V, \\ f(\mathbf{x}) < 0 & \text{for } \mathbf{x} \text{ within } S, \end{cases} \tag{2.3.1}$$

and consider the *Heaviside unit function*

$$H(f) = \begin{cases} 1 & \text{for } \mathbf{x} \text{ in } V, \\ 0 & \text{for } \mathbf{x} \text{ within } S. \end{cases}$$

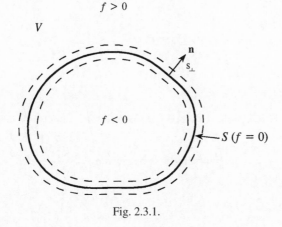

Fig. 2.3.1.

Then, for an arbitrary function $\Phi(\mathbf{x})$ defined in V and on S,

$$\int_{-\infty}^{\infty} \Phi(\mathbf{x}) \nabla H \, d^3\mathbf{x} = \oint_S \Phi(\mathbf{x}) \mathbf{n} \, dS \equiv \oint_S \Phi(\mathbf{x}) \, d\mathbf{S}, \qquad (2.3.2)$$

$$\text{or} \quad \int_{-\infty}^{\infty} \Phi(\mathbf{x}) \frac{\partial H}{\partial x_j} \, d^3\mathbf{x} = \oint_S \Phi(\mathbf{x}) n_j \, dS \equiv \oint_S \Phi(\mathbf{x}) \, dS_j, \qquad (2.3.3)$$

where $H \equiv H(f)$ and $\mathbf{n}$ is the unit normal on S directed into V.

Proof

$$\nabla H(f) \equiv \delta(f) \nabla f \qquad (2.3.4)$$

is nonzero only on S, where ∇f is in the direction of $\mathbf{n}$. The volume integral is therefore confined to the region between the inner and outer faces of a shell of infinitesimal thickness (between the broken line surfaces in Fig. 2.3.1) that just encloses S, and in which the volume element is

$$d^3\mathbf{x} = ds_\perp dS,$$

where $s_\perp = 0$ on S and $s_\perp$ is measured parallel to $\mathbf{n}$. Because $f = 0$ on S we can write, for small values of $s_\perp$,

$$f = \left(\frac{\partial f}{\partial s_\perp} \right)_S s_\perp,$$

where $(\partial f / \partial s_\perp)_S \equiv |\nabla f| > 0$ is evaluated on S.

Therefore,

$$\delta(f) = \delta(|\nabla f| s_\perp) \equiv \frac{\delta(s_\perp)}{|\nabla f|}.$$

Hence,

$$\int_{-\infty}^{\infty} \Phi(\mathbf{x}) \nabla H \, d^3\mathbf{x} \equiv \int_{-\infty}^{\infty} \Phi(\mathbf{x}) \nabla f \delta(f) \, d^3\mathbf{x} = \int_{-\infty}^{\infty} \Phi(\mathbf{x}) \frac{\nabla f}{|\nabla f|} \delta(s_\perp) \, ds_\perp \, dS$$

$$= \oint_S \Phi(\mathbf{x}) \mathbf{n} \, dS, \quad \text{because } \mathbf{n} = \frac{\nabla f}{|\nabla f|}. \qquad \Box$$

2.3.2 Curle's Equation

Curle (1955) has derived a formal solution (called *Curle's equation*) of Lighthill's equation (2.1.12) for the sound produced by turbulence in the vicinity of an arbitrary, *fixed* surface S, defined as above by an equation $f(\mathbf{x}) = 0$ (Fig. 2.3.2). This surface may either enclose a solid body, or merely constitute a control surface used to isolate a fixed region of space containing both solid bodies and fluid or just fluid.

To derive Curle's equation, multiply the momentum equation (2.1.11) by $H \equiv H(f)$, and use the definition (2.1.10) of T_{ij} to obtain

$$\frac{\partial}{\partial t}(\rho v_i H) + \frac{\partial}{\partial x_i}\big(H c_0^2(\rho - \rho_0)\big) = -\frac{\partial}{\partial x_j}(H T_{ij}) + (\rho v_i v_j + p'_{ij})\frac{\partial H}{\partial x_j},$$

$$(2.3.5)$$

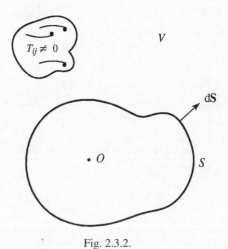

Fig. 2.3.2.

where

$$p'_{ij} = (p - p_0)\delta_{ij} - \sigma_{ij} \qquad (2.3.6)$$

is the *compressive stress* tensor. Repeat this operation for the continuity equation (2.1.8):

$$\frac{\partial}{\partial t}(H(\rho - \rho_0)) + \frac{\partial}{\partial x_i}(H\rho v_i) = (\rho v_i)\frac{\partial H}{\partial x_i}. \qquad (2.3.7)$$

The Formula (2.3.4), $\nabla H = \nabla f \delta(f)$, shows that Equations (2.3.5) and (2.3.7) formally determine the momentum density ρv_i and the density fluctuation $(\rho - \rho_0)$ *in the exterior region* V (where $H(f) \equiv 1$) in terms of the Lighthill stresses T_{ij} in V and sources distributed over the control surface.

An analog of Lighthill's equation (2.1.12) can now be obtained by eliminating $H\rho v_i$ between (2.3.5) and (2.3.7). This is the differential form of Curle's equation

$$\left(\frac{1}{c_0^2}\frac{\partial^2}{\partial t^2} - \nabla^2\right)\left[Hc_0^2(\rho - \rho_0)\right]$$

$$= \frac{\partial^2(HT_{ij})}{\partial x_i \partial x_j} - \frac{\partial}{\partial x_i}\left((\rho v_i v_j + p'_{ij})\frac{\partial H}{\partial x_j}\right) + \frac{\partial}{\partial t}\left(\rho v_j \frac{\partial H}{\partial x_j}\right). \qquad (2.3.8)$$

The equation is valid throughout all space, including the region enclosed by S where $H(f)$ vanishes. The second and third terms on the right-hand side respectively represent *dipole* and *monopole* sources distributed over S. They have the following interpretations:

1. If S is merely an artificial control surface it will enclose fluid, possibly also solid bodies, and may or may not contain turbulence; the surface dipole and monopole sources then represent the influence of this region on the sound radiated in V; in other words the aggregate effect of the dipole and monopole sources accounts for the presence of solid bodies and turbulence within S (when $T_{ij} \neq 0$ in S) and also for the interaction of sound generated outside S with the fluid and solid bodies in S.
2. If S is the boundary of a solid body, the surface dipole represents the production of sound by the unsteady surface force that the body exerts on the exterior fluid, whereas the monopole is responsible for the sound produced by volume pulsations (if any) of the body.

Because Curle's form of Lighthill's equation is valid throughout all space, the outgoing wave solution is found from the general solution (1.6.6) of the wave equation (1.6.3), as before, using the special form (1.7.5) for dipole sources. When account is taken of the transformation formula (2.3.3) this yields **Curle's equation**

$$Hc_0^2(\rho - \rho_0) = \frac{\partial^2}{\partial x_i \partial x_j} \int_V [T_{ij}] \frac{d^3\mathbf{y}}{4\pi|\mathbf{x} - \mathbf{y}|} - \frac{\partial}{\partial x_i} \oint_S [\rho v_i v_j + p'_{ij}] \frac{dS_j(\mathbf{y})}{4\pi|\mathbf{x} - \mathbf{y}|}$$

$$+ \frac{\partial}{\partial t} \oint_S [\rho v_j] \frac{dS_j(\mathbf{y})}{4\pi|\mathbf{x} - \mathbf{y}|}, \qquad (2.3.9)$$

where the square bracket notation such as $[T_{ij}] \equiv T_{ij}(\mathbf{y}, t - |\mathbf{x} - \mathbf{y}|/c_0)$ implies evaluation at the retarded time. Note that, because $H(f) \equiv 0$ inside S, the sum of the three integrals on the right-hand side must also vanish when the field point $\mathbf{x}$ is within S.

2.4 Sound Produced by Turbulence Near a Compact Rigid Body

When the surface S (in Fig. 2.3.2) is *rigid*, Curle's equation (2.3.9) reduces to

$$Hc_0^2(\rho - \rho_0) = \frac{\partial^2}{\partial x_i \partial x_j} \int_V [T_{ij}] \frac{d^3\mathbf{y}}{4\pi|\mathbf{x} - \mathbf{y}|} - \frac{\partial}{\partial x_i} \oint_S [p'_{ij}] \frac{dS_j(\mathbf{y})}{4\pi|\mathbf{x} - \mathbf{y}|}. \quad (2.4.1)$$

We now use this solution to determine the order of magnitude of the sound generated by an acoustically *compact* body within a turbulent flow. Compactness usually requires the Mach number $M \sim v/c_0 \ll 1$, and we shall assume this to be the case in the following.

The contribution from the quadrupole integral in (2.4.1) is estimated as in Section 2.2. To deal with the surface dipole, note first that for turbulence of velocity v and correlation scale ℓ, the orders of magnitude of the pressure and viscous components of the compressive stress tensor

$$p'_{ij} = (p - p_0)\delta_{ij} - \sigma_{ij}$$

are

$$(p - p_0) \sim \rho_0 v^2, \qquad \sigma \sim \eta \frac{v}{\ell};$$

that is,

$$\frac{(p - p_0)}{\sigma} \sim \frac{\rho_0 v\ell}{\eta} = \frac{v\ell}{\nu},$$

where $\nu = \eta/\rho_0$ is the kinematic viscosity. The dimensionless ratio $\text{Re} = v\ell/\nu$ is the *Reynolds number* and is always very large ($\sim 10^4$ or more) in turbulent flow. This means that viscous contributions to the surface force can be neglected.

In the far field the pressure $p(\mathbf{x}, t) = c_0^2(\rho - \rho_0)(\mathbf{x}, t)$, and $H(f) = 1$. Thus, applying the far-field dipole approximation (1.9.6), and neglecting retarded time variations $\mathbf{x} \cdot \mathbf{y}/c_0|\mathbf{x}|$ because S is compact, the dipole sound pressure p_d, say, can be written

$$p_\mathrm{d} \approx \frac{x_i}{4\pi c_0|\mathbf{x}|^2} \frac{\partial}{\partial t} \oint_S (p - p_0)\left(\mathbf{y}, t - \frac{|\mathbf{x}|}{c_0}\right) dS_i = \frac{x_i}{4\pi c_0|\mathbf{x}|^2} \frac{dF_i}{dt}\left(t - \frac{|\mathbf{x}|}{c_0}\right),$$

$$|\mathbf{x}| \to \infty, \quad (2.4.2)$$

where $\mathbf{F}(t)$ is the unsteady force exerted on the fluid by the body. The contribution to p_d from a surface element of diameter ℓ within which the turbulence surface pressure fluctuations are correlated is evidently of order

$$\frac{1}{c_0|\mathbf{x}|} \frac{v}{\ell} \times (\rho_0 v^2 \ell^2) = \frac{\ell}{|\mathbf{x}|} \rho_0 v^2 M,$$

which exceeds by an order of magnitude ($1/M \gg 1$) the sound pressure (2.2.4) produced by a quadrupole in V of length scale ℓ. If A is the total surface area wetted by the turbulent flow, there are A/ℓ^2 independently radiating surface elements, and the total power radiated by the dipoles is

$$\Pi_\mathrm{d} \sim 4\pi |\mathbf{x}|^2 \times \left(\frac{p_d^2}{\rho_0 c_0}\right) \sim A\rho_0 v^3 M^3. \quad (2.4.3)$$

The direct power radiated by quadrupoles occupying a volume V_0 is $\Pi_q \sim (V_0/\ell)\rho_0 v^3 M^5$, the same as in the absence of the body (see Section 2.2). The sound produced by the turbulence near S is therefore dominated by the dipole when M is small, and as $M \to 0$ the acoustic power exceeds the quadrupole power by a factor $\sim 1/M^2 \gg 1$. Precisely how small M should be for this to be true depends on the details of the flow, which determine the appropriate values of A and V_0/ℓ.

This increase in acoustic efficiency brought about by surface dipoles on an acoustically compact body occurs also for arbitrary, *noncompact* bodies when turbulence interacts with compact structural elements, such as edges, corners, and protuberances.

2.5 Radiation from a Noncompact Surface

Consider a compact turbulent eddy in $x_2 > 0$ adjacent to an infinite, plane rigid wall at $x_2 = 0$ (Fig. 2.5.1). Let us apply the rigid surface form (2.4.1) of Curle's equation to calculate the radiation. At high Reynolds numbers and at $\mathbf{x}$ in the

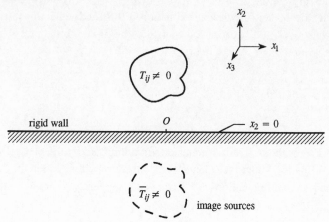

observer • x

$T_{ij} \neq 0$

rigid wall   O $x_2 = 0$

$\overline{T}_{ij} \neq 0$

image sources

image point • $\overline{x}$

Fig. 2.5.1.

acoustic far field (where $p(\mathbf{x}, t) \equiv c_0^2(\rho - \rho_0)$ and $H(f) \equiv H(x_2) = 1$) we find

$$
p(\mathbf{x}, t) \approx \frac{x_i x_j}{4\pi c_0^2 |\mathbf{x}|^3} \frac{\partial^2}{\partial t^2} \int T_{ij}\left(\mathbf{y}, t - \frac{|\mathbf{x}|}{c_0}\right) d^3\mathbf{y}
$$
$$
+ \frac{x_2}{4\pi c_0 |\mathbf{x}|^2} \frac{\partial}{\partial t} \oint_{y_2=0} (p - p_0)\left(\mathbf{y}, t - \frac{|\mathbf{x}|}{c_0} + \frac{x_1 y_1 + x_3 y_3}{c_0 |\mathbf{x}|}\right) dy_1 \, dy_3,
$$
$$
|\mathbf{x}| \to \infty. \quad (2.5.1)
$$

Retarded time variations have been neglected in the integral over the volume of the compact turbulent eddy. We have not done this in the surface pressure integral, because this tends to extend over a larger region than the Reynolds stress fluctuations responsible for it (indeed, the *acoustic* component of the pressure, as opposed to the near field *hydrodynamic pressure,* extends out to infinity on the wall, decaying only very slowly like $1/|\mathbf{x}|$).

The value of the surface integral cannot be estimated by a naive order-of-magnitude calculation of the kind performed in Section 2.4 for a compact body, because for an infinite plane wall the domain of integration includes the acoustic region, and therefore involves an unknown and possibly important contribution from the acoustic pressure that we are trying to calculate! The difficulty was resolved by Powell (1960) by the ingenious device of applying Curle's solution

(2.4.1) at the *image* $\bar{\mathbf{x}} = (x_1, -x_2, x_3)$ in the wall of the far field observation point $\mathbf{x}$. At the image point $H(f) = H(x_2) \equiv 0$, and therefore

$$0 \approx \frac{\bar{x}_i \bar{x}_j}{4\pi c_0^2 |\mathbf{x}|^3} \frac{\partial^2}{\partial t^2} \int T_{ij} \left(\mathbf{y}, t - \frac{|\mathbf{x}|}{c_0} \right) d^3 \mathbf{y}$$

$$- \frac{x_2}{4\pi c_0 |\mathbf{x}|^2} \frac{\partial}{\partial t} \oint_{y_2=0} (p - p_0) \left(\mathbf{y}, t - \frac{|\mathbf{x}|}{c_0} + \frac{x_1 y_1 + x_3 y_3}{c_0 |\mathbf{x}|} \right) dy_1 \, dy_3,$$

$$|\mathbf{x}| \to \infty. \quad (2.5.2)$$

The surface integral term in this formula is equal but opposite in sign to that in the original solution (2.5.1), which is now seen to exactly represent the quadrupole sound generated by a system of *image quadrupoles* in the wall! Adding (2.5.1) and (2.5.2), we find

$$p(\mathbf{x}, t) \approx \frac{(x_i x_j + \bar{x}_i \bar{x}_j)}{4\pi c_0^2 |\mathbf{x}|^3} \frac{\partial^2}{\partial t^2} \int T_{ij} \left(\mathbf{y}, t - \frac{|\mathbf{x}|}{c_0} \right) d^3 \mathbf{y}$$

$$\approx \frac{(x_i x_j + \bar{x}_i \bar{x}_j)}{4\pi c_0^2 |\mathbf{x}|^3} \frac{\partial^2}{\partial t^2} \int \rho_0 v_i v_j \left(\mathbf{y}, t - \frac{|\mathbf{x}|}{c_0} \right) d^3 \mathbf{y}, \quad |\mathbf{x}| \to \infty. \quad (2.5.3)$$

Therefore, the apparently strong contribution from the surface pressure dipoles actually integrates to a term of quadrupole strength. This is a consequence of the *Kraichnan–Phillips* theorem (see Howe, 1998a), according to which the net unsteady component of the normal force between an *infinite* plane wall and an incompressible fluid must vanish identically

$$\oint_{y_2=0} (p - p_0)(\mathbf{y}, t) \, dy_1 \, dy_3 \equiv 0. \quad (2.5.4)$$

Thus, extreme care must be exercised when using Curle's equation to estimate the sound produced by turbulence interacting with large surfaces. As a general rule, the surface contribution will be comparable to that from the turbulence quadrupoles whenever *the characteristic wavelength of the sound is smaller than the radius of curvature of the surface.*

Problems 2

1. Show that the acoustic efficiency of a compact sphere of radius R executing small amplitude translational oscillations at velocity $U = U_0 \sin(\omega t)$ is

$$\frac{\Pi_a}{\Pi_0} \sim \left(\frac{\omega R}{c_0} \right)^3,$$

where

$$\Pi_a = \frac{\pi \omega R^3}{6} \rho_0 U_0^2 \left(\frac{\omega R}{c_0} \right)^3, \qquad \Pi_0 = \frac{2\omega R^3}{3} \rho_0 U_0^2$$

are respectively the average acoustic and hydrodynamic powers fed into the fluid during the *quarter* cycle $0 < \omega t < \pi/2$.

Explain the significance of averaging only over $0 < \omega t < \pi/2$.

2. What is the efficiency in Problem 1 when the sphere pulsates with small amplitude normal velocity $v_n = U_0 \sin(\omega t)$?

3. The wake behind a bluff body fixed in a nominally steady, low Mach number flow at speed U produces a drag force equal to $C_D A \frac{1}{2} \rho_0 U^2$, where C_D is the *drag coefficient* (which may be regarded as constant), and A is the projected cross-sectional area of the body in the flow direction. Derive an approximate formula for the far-field acoustic pressure radiated by the body when U contains a small amplitude, time-harmonic component such that $U = U_0 + u \cos(\omega t)$, where U_0 and u are constant and $u \ll U_0$.

4. Show that Powell's solution (2.5.3) for the sound generated by turbulence adjacent to a rigid plane wall is identical with the solution of Curle's differential equation (2.3.8) determined by the modified Green's function

$$G(\mathbf{x}, \mathbf{y}, t - \tau) = \frac{1}{4\pi |\mathbf{x} - \mathbf{y}|} \delta \left(t - \tau - \frac{|\mathbf{x} - \mathbf{y}|}{c_0} \right)$$
$$+ \frac{1}{4\pi |\bar{\mathbf{x}} - \mathbf{y}|} \delta \left(t - \tau - \frac{|\bar{\mathbf{x}} - \mathbf{y}|}{c_0} \right),$$

where $\bar{\mathbf{x}} = (x_1, -x_2, x_3)$. $G(\mathbf{x}, \mathbf{y}, t - \tau)$ is the solution of what problem of linear acoustics?

3

The Compact Green's Function

3.1 The Influence of Solid Boundaries

Let us return to the general problem of linear acoustics. To fix ideas, we shall frame the present discussion in terms of Equation (1.3.9)

$$\left(\frac{1}{c_0^2} \frac{\partial^2}{\partial t^2} - \nabla^2 \right) \varphi = -q(\mathbf{x}, t) \tag{3.1.1}$$

for the velocity potential, but our conclusions will be applicable quite generally. This equation determines φ in terms of a specified source distribution $q(\mathbf{x}, t)$. In the absence of solid boundaries (in *free space*) the results of Section 1.6 enable us to represent φ in the form

$$\varphi(\mathbf{x}, t) = \iint_{-\infty}^{\infty} -q(\mathbf{y}, \tau) G(\mathbf{x}, \mathbf{y}, t - \tau) \, d^3\mathbf{y} \, d\tau, \tag{3.1.2}$$

where $G(\mathbf{x}, \mathbf{y}, t - \tau)$ is the free space Green's function

$$G(\mathbf{x}, \mathbf{y}, t - \tau) = \frac{1}{4\pi |\mathbf{x} - \mathbf{y}|} \delta\left(t - \tau - \frac{|\mathbf{x} - \mathbf{y}|}{c_0} \right), \tag{3.1.3}$$

that is, $G(\mathbf{x}, \mathbf{y}, t - \tau)$ is the outgoing wave solution of

$$\left(\frac{1}{c_0^2} \frac{\partial^2}{\partial t^2} - \nabla^2 \right) G = \delta(\mathbf{x} - \mathbf{y})\delta(t - \tau), \quad \text{where } G = 0 \quad \text{for } t < \tau. \tag{3.1.4}$$

In our discussion of Curle's extension of Lighthill's theory in Chapter 2, it was found that the presence of a solid boundary S in the vicinity of the turbulence quadrupole sources T_{ij} resulted in the appearance of additional dipole and monopole sources distributed over S, represented by the second and third terms on the right-hand side of Equation (2.3.8). Curle's solution (2.3.9) of this

41

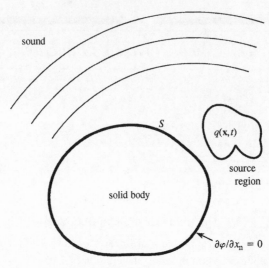

Fig. 3.1.1.

equation was derived by using the retarded potential formula (1.6.6), obtained by use of the free space Green's function (3.1.3). A similar representation involving surface distributions of dipoles and monopoles is obtained for φ when we attempt to solve Equation (3.1.1) using Green's function (3.1.3) in the situation illustrated in Fig. 3.1.1, where the source distribution $q(\mathbf{x}, t)$ is adjacent to a rigid boundary S on which the normal derivative $\partial \varphi / \partial x_n = 0$.

It would be very convenient if we could modify the functional form of $G(\mathbf{x}, \mathbf{y}, t - \tau)$ so that it *automatically* takes account of the contributions from the dipole and monopole sources on S, inasmuch that no surface integrals occur in the final solution. To do this, we must find a solution of Green's function equation (3.1.4) that satisfies appropriate *boundary conditions* on S. The solution φ of (3.1.1) is then once again given by Formula (3.1.2) in terms of the *modified* Green's function, there being no additional surface integrals to evaluate.

The main practical difficulty is the calculation of the modified Green's function. Although it is always possible in principle, exact analytical representations are known only for solid bodies of very simple shapes (such as spheres, circular cylinders, and half-planes). However, it turns out that a relatively simple and general approximate formula can be found for the modified Green's function for those problems where it is known that the typical wavelength of the sound produced by the source distribution $q(\mathbf{x}, t)$ is *large compared to one or more principal dimensions* of the solid body S. This is called the *compact Green's function*.

To simplify the calculation of the compact Green's function, we use the Fourier integral formula

$$\delta(t - \tau) = \frac{1}{2\pi} \int_{-\infty}^{\infty} e^{-i\omega(t-\tau)} \, d\omega, \qquad (3.1.5)$$

which expresses the δ function as a linear combination of time-harmonic oscillations of frequency ω. The formula is proved by observing that no real system can oscillate at infinitely large frequencies, and therefore in all practical problems $e^{-i\omega(t-\tau)}$ can be replaced by $e^{-i\omega(t-\tau)-\epsilon|\omega|}$ for arbitrarily small $\epsilon > 0$. Then,

$$\frac{1}{2\pi} \int_{-\infty}^{\infty} e^{-i\omega(t-\tau)} d\omega \equiv \lim_{\epsilon \to +0} \frac{1}{2\pi} \int_{-\infty}^{\infty} e^{-i\omega(t-\tau)-\epsilon|\omega|} \, d\omega$$

$$= \lim_{\epsilon \to +0} \frac{\epsilon}{\pi[\epsilon^2 + (t-\tau)^2]}.$$

The final term on the right is the usual definition of $\delta(t - \tau)$ as the limit of an 'ϵ-sequence' (Lighthill, 1958).

If we now put

$$G(\mathbf{x}, \mathbf{y}, t - \tau) = \frac{-1}{2\pi} \int_{-\infty}^{\infty} \hat{G}(\mathbf{x}, \mathbf{y}, \omega) e^{-i\omega(t-\tau)} \, d\omega, \qquad (3.1.6)$$

then the substitution of this and (3.1.5) into the Green's function equation (3.1.4) shows that, for each frequency ω, $\hat{G}(\mathbf{x}, \mathbf{y}, \omega)$ is the solution of

$$\left(\nabla^2 + \kappa_0^2\right)\hat{G}(\mathbf{x}, \mathbf{y}, \omega) = \delta(\mathbf{x} - \mathbf{y}), \qquad (3.1.7)$$

where $\kappa_0 = \omega/c_0$ is called the **acoustic wave number.** Sound of frequency ω has wavelength

$$\lambda = \frac{2\pi}{\kappa_0}.$$

Thus, a solid body of characteristic dimension ℓ is compact for waves of frequency ω provided that

$$\frac{\ell}{\lambda} = \frac{\kappa_0 \ell}{2\pi} \ll 1. \qquad (3.1.8)$$

This condition will be used below in Section 3.4 to calculate the compact Green's function.

3.2 The Helmholtz Equation

The equations

$$\left(\nabla^2 + \kappa_0^2\right)\hat{\varphi} = 0, \qquad \left(\nabla^2 + \kappa_0^2\right)\hat{\varphi} = \hat{q}(\mathbf{x}, \omega) \tag{3.2.1}$$

are known respectively as the Helmholtz equation and the *inhomogeneous* Helmholtz equation. The source term $\hat{q}(\mathbf{x}, \omega)$ represents one frequency component of the source $q(\mathbf{x}, t)$ of Equation (3.1.1), so that

$$q(\mathbf{x}, t) = \int_{-\infty}^{\infty} \hat{q}(\mathbf{x}, \omega)e^{-i\omega t}\, d\omega. \tag{3.2.2}$$

Therefore, because (differentiating under the integral sign)

$$\frac{1}{c_0^2}\frac{\partial^2}{\partial t^2}\int_{-\infty}^{\infty} \hat{\varphi}(\mathbf{x}, \omega)e^{-i\omega t}\, d\omega = \frac{1}{c_0^2}\int_{-\infty}^{\infty} -\omega^2\hat{\varphi}(\mathbf{x}, \omega)e^{-i\omega t}\, d\omega$$

$$\equiv -\int_{-\infty}^{\infty} \kappa_0^2\hat{\varphi}(\mathbf{x}, \omega)e^{-i\omega t}\, d\omega,$$

the solution $\hat{\varphi}(\mathbf{x}, \omega)$ of the inhomogeneous equation is related to the solution of (3.1.1) by

$$\varphi(\mathbf{x}, t) = \int_{-\infty}^{\infty} \hat{\varphi}(\mathbf{x}, \omega)e^{-i\omega t}\, d\omega. \tag{3.2.3}$$

3.2.1 The Point Source

Equation (3.1.7) determines Green's function $\hat{G}(\mathbf{x}, \mathbf{y}, \omega)$ for the inhomogeneous Helmholtz equation. The *free space* Green's function can be found by the method used in Section 1.5 for the wave equation. If we temporarily set $\mathbf{y} = 0$, we have to find the radially symmetric solution of

$$\left(\nabla^2 + \kappa_0^2\right)\hat{G} = \delta(\mathbf{x}). \tag{3.2.4}$$

In the usual way (see Section 1.5), we have

$$\frac{\partial^2}{\partial r^2}(r\hat{G}) + \kappa_0^2(r\hat{G}) = 0 \qquad \text{for } r = |\mathbf{x}| > 0,$$

and therefore

$$\hat{G} = \frac{Ae^{i\kappa_0 r}}{r} + \frac{Be^{-i\kappa_0 r}}{r}, \tag{3.2.5}$$

where A, B remain to be determined.

To do this recall that our solution represents one component of a time-dependent acoustic problem of frequency ω. Since the time factor is $e^{-i\omega t}$, the two terms on the right-hand side of (3.2.5) correspond to propagating sound waves of the form

$$\frac{Ae^{-i\omega\left(t-\frac{r}{c_0}\right)}}{r} + \frac{Be^{-i\omega\left(t+\frac{r}{c_0}\right)}}{r},$$

the second of which represents waves converging on the source from infinity, and must therefore be rejected because of the radiation condition. Hence, $B = 0$. The value of the remaining constant A is found by extending the solution to include the region occupied by the source at $r = 0$ by writing

$$\hat{G} = \lim_{\epsilon \to 0} \frac{Ae^{i\kappa_0 r}}{(r^2 + \epsilon^2)^{\frac{1}{2}}}.$$

By substituting the solution into (3.2.4) and using the Formula (1.4.8) we find $A = -1/4\pi$.

The free space Green's function for the inhomogeneous Helmholtz equation (the solution of (3.1.7)) is now obtained by replacing $r = |\mathbf{x}|$ by $|\mathbf{x} - \mathbf{y}|$

$$\hat{G}(\mathbf{x}, \mathbf{y}, \omega) = \frac{-e^{i\kappa_0|\mathbf{x}-\mathbf{y}|}}{4\pi|\mathbf{x} - \mathbf{y}|}. \tag{3.2.6}$$

Because the source $\hat{q}(\mathbf{x}, \omega)$ in the second of Equations (3.2.1) can be expressed as a superposition of point sources by means of

$$\hat{q}(\mathbf{x}, \omega) = \int_{-\infty}^{\infty} \hat{q}(\mathbf{y}, \omega)\delta(\mathbf{x} - \mathbf{y}) \, d^3\mathbf{y},$$

the solution of the inhomogeneous Helmholtz equation in an unbounded medium can be written

$$\varphi(\mathbf{x}, \omega) = \int_{-\infty}^{\infty} \hat{G}(\mathbf{x}, \mathbf{y}, \omega)\hat{q}(\mathbf{y}, \omega) \, d^3\mathbf{y} \equiv \frac{-1}{4\pi} \int_{-\infty}^{\infty} \frac{\hat{q}(\mathbf{y}, \omega)e^{i\kappa_0|\mathbf{x}-\mathbf{y}|}}{|\mathbf{x} - \mathbf{y}|} \, d^3\mathbf{y}. \tag{3.2.7}$$

3.2.2 Dipole and Quadrupole Sources

The method of integration by parts described in Section 1.7 can be used to show that the corresponding solutions for the dipole and quadrupole sources

$$\hat{q}(\mathbf{x}, \omega) = \frac{\partial f_j}{\partial x_j}(\mathbf{x}, \omega) \quad \text{and} \quad \hat{q}(\mathbf{x}, \omega) = \frac{\partial^2 T_{ij}}{\partial x_i \partial x_j}(\mathbf{x}, \omega) \tag{3.2.8}$$

are respectively

$$\hat{\varphi}(\mathbf{x}, \omega) = \frac{-1}{4\pi} \frac{\partial}{\partial x_j} \int_{-\infty}^{\infty} \frac{f_j(\mathbf{y}, \omega)e^{i\kappa_0|\mathbf{x}-\mathbf{y}|}}{|\mathbf{x} - \mathbf{y}|} d^3\mathbf{y},$$

and

$$\hat{\varphi}(\mathbf{x}, \omega) = \frac{-1}{4\pi} \frac{\partial^2}{\partial x_i \partial x_j} \int_{-\infty}^{\infty} \frac{T_{ij}(\mathbf{y}, \omega)e^{i\kappa_0|\mathbf{x}-\mathbf{y}|}}{|\mathbf{x} - \mathbf{y}|} d^3\mathbf{y}. \qquad (3.2.9)$$

Example For a point dipole at the origin orientated in the x_1 direction

$$\hat{q}(\mathbf{x}, \omega) = \frac{\partial}{\partial x_1}(f_1\delta(\mathbf{x})).$$

Therefore,

$$\hat{\varphi}(\mathbf{x}, \omega) = \frac{-1}{4\pi} \frac{\partial}{\partial x_1} \int_{-\infty}^{\infty} \frac{f_1\delta(\mathbf{y})e^{i\kappa_0|\mathbf{x}-\mathbf{y}|}}{|\mathbf{x} - \mathbf{y}|} d^3\mathbf{y} \approx \frac{-i\kappa_0 x_1 f_1 e^{i\kappa_0|\mathbf{x}|}}{4\pi|\mathbf{x}|^2}, \quad |\mathbf{x}| \to \infty.$$
$$(3.2.10)$$

3.2.3 Green's Function for the Wave Equation

Let us verify the general relation (3.1.6) between the Green's functions $G(\mathbf{x}, \mathbf{y}, t-\tau)$ and $\hat{G}(\mathbf{x}, \mathbf{y}, \omega)$, respectively, for the wave equation and the inhomogeneous Helmholtz equation in the special case in which there are no solid boundaries. According to this formula we find, using the expression (3.2.6) for the free space Green's function $\hat{G}$,

$$\begin{aligned}
G(\mathbf{x}, \mathbf{y}, t - \tau) &= \frac{-1}{2\pi} \int_{-\infty}^{\infty} \hat{G}(\mathbf{x}, \mathbf{y}, \omega)e^{-i\omega(t-\tau)} d\omega \\
&= \frac{1}{8\pi^2|\mathbf{x} - \mathbf{y}|} \int_{-\infty}^{\infty} e^{-i\omega\left(t-\tau-\frac{|\mathbf{x}-\mathbf{y}|}{c_0}\right)} d\omega \\
&= \frac{1}{4\pi|\mathbf{x} - \mathbf{y}|}\delta\left(t - \tau - \frac{|\mathbf{x} - \mathbf{y}|}{c_0}\right),
\end{aligned}$$

which is precisely Equation (3.1.3).

3.3 The Reciprocal Theorem

The calculation of the compact Green's function is greatly simplified by application of the *reciprocal theorem*. We need to consider only a special case of

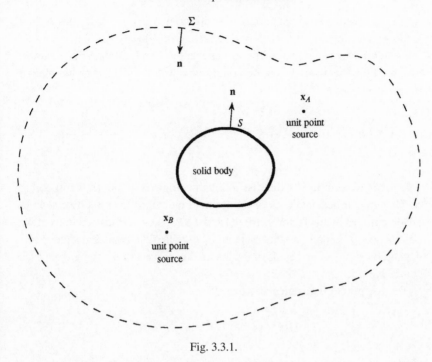

Fig. 3.3.1.

this very general theorem of mechanics, which was first used with great effect in acoustics by Lord Rayleigh (1945).

Consider the two acoustic problems indicated in Fig. 3.3.1, in which sound of frequency ω is generated by two unit point sources at $\mathbf{x} = \mathbf{x}_A$ and $\mathbf{x} = \mathbf{x}_B$ in the presence of a solid body S. We denote the functional forms of the respective velocity potentials generated by these sources by $\hat{G}(\mathbf{x}, \mathbf{x}_A, \omega)$ and $\hat{G}(\mathbf{x}, \mathbf{x}_B, \omega)$, where

$$\left(\nabla^2 + \kappa_0^2\right)\hat{G}(\mathbf{x}, \mathbf{x}_A, \omega) = \delta(\mathbf{x} - \mathbf{x}_A), \qquad (3.3.1)$$

$$\left(\nabla^2 + \kappa_0^2\right)\hat{G}(\mathbf{x}, \mathbf{x}_B, \omega) = \delta(\mathbf{x} - \mathbf{x}_B). \qquad (3.3.2)$$

In addition $\hat{G}(\mathbf{x}, \mathbf{x}_A, \omega)$ and $\hat{G}(\mathbf{x}, \mathbf{x}_B, \omega)$ must satisfy appropriate mechanical boundary conditions on S. We take these to have the same general linear form

$$\frac{\partial \hat{G}}{\partial x_\mathrm{n}}(\mathbf{x}, \mathbf{x}_A, \omega) = \frac{\hat{G}(\mathbf{x}, \mathbf{x}_A, \omega)}{\mathcal{Z}(\mathbf{x}, \omega)}, \qquad \frac{\partial \hat{G}}{\partial x_n}(\mathbf{x}, \mathbf{x}_B, \omega) = \frac{\hat{G}(\mathbf{x}, \mathbf{x}_B, \omega)}{\mathcal{Z}(\mathbf{x}, \omega)},$$

$$\text{for } \mathbf{x} \text{ on } S, \quad (3.3.3)$$

where x_n is measured in the normal direction from S *into the fluid* and $\mathcal{Z}(\mathbf{x}, \omega)$ is the surface impedance. For a *rigid* surface, $\mathcal{Z}(\mathbf{x}, \omega) = \infty$.

At large distances from S, in the acoustic far field, both solutions are assumed to exhibit the characteristics of outgoing sound waves, such that (with implicit time dependence $e^{-i\omega t}$)

$$\hat{G}(\mathbf{x}, \mathbf{x}_A, \omega) \sim \frac{f_A(\theta, \phi)e^{i\kappa_0 r}}{r}, \qquad \hat{G}(\mathbf{x}, \mathbf{x}_B, \omega) \sim \frac{f_B(\theta, \phi)e^{i\kappa_0 r}}{r},$$

$$r \equiv |\mathbf{x}| \to \infty, \quad (3.3.4)$$

where it may be supposed that the coordinate origin is in the neighborhood of S. The angular dependencies of the far-field radiations from the two sources are determined by the factors $f_A(\theta, \phi)$ and $f_B(\theta, \phi)$, which are functions of the polar angles θ, ϕ defining the orientation of the far field point $\mathbf{x}$, and generally depend strongly on the details of the interaction of the volume flows from each source with S.

The **reciprocal theorem** states that

$$\hat{G}(\mathbf{x}_A, \mathbf{x}_B, \omega) = \hat{G}(\mathbf{x}_B, \mathbf{x}_A, \omega). \quad (3.3.5)$$

That is, the potential at $\mathbf{x}_A$ produced by the point source at $\mathbf{x}_B$ is equal to the potential at $\mathbf{x}_B$ produced by an equal point source at $\mathbf{x}_A$.

Proof. Multiply Equation (3.3.1) by $\hat{G}(\mathbf{x}, \mathbf{x}_B, \omega)$ and Equation (3.3.2) by $\hat{G}(\mathbf{x}, \mathbf{x}_A, \omega)$, subtract the resulting equations and integrate over the volume bounded by the surface S and by a large surface Σ in the acoustic far field. Green's identity

$$\hat{G}(\mathbf{x}, \mathbf{x}_B, \omega)\nabla^2\hat{G}(\mathbf{x}, \mathbf{x}_A, \omega) - \hat{G}(\mathbf{x}, \mathbf{x}_A, \omega)\nabla^2\hat{G}(\mathbf{x}, \mathbf{x}_B, \omega)$$

$$= \mathrm{div}(\hat{G}(\mathbf{x}, \mathbf{x}_B, \omega)\nabla\hat{G}(\mathbf{x}, \mathbf{x}_A, \omega) - \hat{G}(\mathbf{x}, \mathbf{x}_A, \omega)\nabla\hat{G}(\mathbf{x}, \mathbf{x}_B, \omega)),$$

and the divergence theorem permit the volume integral of the term obtained from the left-hand sides to be expressed as surface integrals over S and Σ, whereas the integrals involving the δ functions can be evaluated explicitly. This procedure gives

$$\oint_{S+\Sigma} \left(\hat{G}(\mathbf{x}, \mathbf{x}_A, \omega)\frac{\partial \hat{G}}{\partial x_n}(\mathbf{x}, \mathbf{x}_B, \omega) - \hat{G}(\mathbf{x}, \mathbf{x}_B, \omega)\frac{\partial \hat{G}}{\partial x_n}(\mathbf{x}, \mathbf{x}_A, \omega) \right) dS$$

$$= \hat{G}(\mathbf{x}_B, \mathbf{x}_A, \omega) - \hat{G}(\mathbf{x}_A, \mathbf{x}_B, \omega).$$

The surface integral over S vanishes because of the impedance conditions (3.3.3). The surface integral over Σ vanishes because of conditions (3.3.4) and because $\partial\theta/\partial x_n$ and $\partial\phi/\partial x_n$ are each of order $1/r$ as $r \to \infty$, and therefore

$$\frac{\partial\hat{G}}{\partial x_n}(\mathbf{x}, \mathbf{x}_A, \omega) \sim f_A(\theta, \phi)\frac{i\kappa_0 e^{i\kappa_0 r}}{r}\frac{\partial r}{\partial x_n},$$

$$\frac{\partial\hat{G}}{\partial x_n}(\mathbf{x}, \mathbf{x}_B, \omega) \sim f_B(\theta, \phi)\frac{i\kappa_0 e^{i\kappa_0 r}}{r}\frac{\partial r}{\partial x_n}, \quad \text{as } r \to \infty.$$

This proves the theorem.

The result is usually expressed as the simple reciprocal relation

$$\hat{G}(\mathbf{x}, \mathbf{y}, \omega) = \hat{G}(\mathbf{y}, \mathbf{x}, \omega). \tag{3.3.6}$$

$\square$

3.4 Time-Harmonic Compact Green's Function

We are now ready to derive the compact Green's function $\hat{G}(\mathbf{x}, \mathbf{y}, \omega)$ for the problem depicted in Fig. 3.4.1. We have to solve

$$\left(\nabla^2 + \kappa_0^2\right)\hat{G}(\mathbf{x}, \mathbf{y}, \omega) = \delta(\mathbf{x} - \mathbf{y}), \qquad \frac{\partial\hat{G}}{\partial x_n} = 0 \quad \text{on } S, \tag{3.4.1}$$

where the rigid body S is assumed to be *acoustically compact*. The influence of

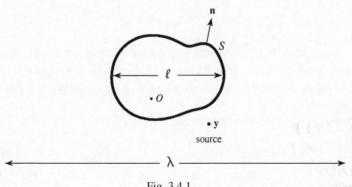

Fig. 3.4.1.

a solid body on the production of sound by neighboring sources is equivalent to an additional distribution of monopoles and dipoles on S. The compact Green's function includes a first approximation for the net effect of these monopole and dipole distributions, obviating the need to evaluate surface integrals.

In practice, we are interested primarily in calculating the sound in the far field of the body. Let ℓ denote the characteristic diameter of the body, and take the coordinate origin at O within S. The source point $\mathbf{y}$ is assumed to be close to S (so that $|\mathbf{y}| \sim \ell$) and the observer at $\mathbf{x}$ is taken to be in the *acoustic far field*. The compactness condition (3.1.8) therefore implies that

$$\kappa_0 \ell \ll 1 \quad \text{and} \quad \kappa_0 |\mathbf{y}| \ll 1.$$

In these circumstances the compact approximation for $\hat{G}(\mathbf{x}, \mathbf{y}, \omega)$ can be found very easily from the solution of the *reciprocal problem:*

$$\left(\frac{\partial^2}{\partial y_1^2} + \frac{\partial^2}{\partial y_2^2} + \frac{\partial^2}{\partial y_3^2} + \kappa_0^2 \right) \hat{G}(\mathbf{y}, \mathbf{x}, \omega) = \delta(\mathbf{y} - \mathbf{x}), \qquad \frac{\partial \hat{G}}{\partial y_n} = 0 \quad \text{on } S,$$
$$(3.4.2)$$

where the source is at the far-field point $\mathbf{x}$, and $\hat{G}(\mathbf{y}, \mathbf{x}, \omega)$ is determined as a function of $\mathbf{y}$ close to S. The solution of (3.4.1) is then given by the reciprocal theorem (Section 3.3) $\hat{G}(\mathbf{x}, \mathbf{y}, \omega) = \hat{G}(\mathbf{y}, \mathbf{x}, \omega)$ (the potential $\hat{G}(\mathbf{x}, \mathbf{y}, \omega)$ at the far-field point $\mathbf{x}$ produced by the point source at $\mathbf{y}$ is exactly equal to the potential $\hat{G}(\mathbf{y}, \mathbf{x}, \omega)$ produced at the near-field point $\mathbf{y}$ by an equal point source at the far-field point $\mathbf{x}$).

To solve (3.4.2), we put

$$\hat{G}(\mathbf{y}, \mathbf{x}, \omega) = \hat{G}_0(\mathbf{y}, \mathbf{x}, \omega) + \hat{G}'(\mathbf{y}, \mathbf{x}, \omega)$$
$$\equiv \frac{-e^{i\kappa_0 |\mathbf{x} - \mathbf{y}|}}{4\pi |\mathbf{x} - \mathbf{y}|} + \hat{G}'(\mathbf{y}, \mathbf{x}, \omega)$$

where $\hat{G}_0(\mathbf{y}, \mathbf{x}, \omega)$ is the spherically spreading wave generated by the point source at $\mathbf{x}$ when the presence of the solid is ignored. The term $\hat{G}'(\mathbf{y}, \mathbf{x}, \omega)$ is the velocity potential of the motion produced in the fluid when this wave impinges on S.

When $|\mathbf{x}| \to \infty$, the approximations

$$|\mathbf{x} - \mathbf{y}| \approx |\mathbf{x}| - \frac{\mathbf{x} \cdot \mathbf{y}}{|\mathbf{x}|} \equiv |\mathbf{x}| - \frac{x_j y_j}{|\mathbf{x}|} \quad \text{and} \quad \frac{1}{|\mathbf{x} - \mathbf{y}|} \approx \frac{1}{|\mathbf{x}|} + \frac{\mathbf{x} \cdot \mathbf{y}}{|\mathbf{x}|^3} \approx \frac{1}{|\mathbf{x}|}$$

and the condition $\kappa_0 y_j \sim \kappa_0 \ell \ll 1$ imply that

$$\hat{G}_0(\mathbf{y}, \mathbf{x}, \omega) \equiv \frac{-e^{i\kappa_0|\mathbf{x}-\mathbf{y}|}}{4\pi|\mathbf{x}-\mathbf{y}|} \approx \frac{-e^{i\kappa_0|\mathbf{x}|}}{4\pi|\mathbf{x}|} \times e^{-\frac{i\kappa_0 x_j y_j}{|\mathbf{x}|}}$$

$$\approx \frac{-e^{i\kappa_0|\mathbf{x}|}}{4\pi|\mathbf{x}|}\left(1 - \frac{i\kappa_0 x_j y_j}{|\mathbf{x}|} + \mathrm{O}(\kappa_0\ell)^2\right) \qquad (3.4.3)$$

The linear dependence on y_j in the second line of this formula represents the first approximation (of order $\kappa_0\ell$) in a power series expansion of rapidly decreasing terms that describes the variation of the incident spherical wave close to the body. Thus, regarded as a function of $\mathbf{y}$, the terms shown explicitly in

$$\hat{G}_0(\mathbf{y}, \mathbf{x}, \omega) = \frac{-e^{i\kappa_0|\mathbf{x}|}}{4\pi|\mathbf{x}|} + \frac{e^{i\kappa_0|\mathbf{x}|}}{4\pi|\mathbf{x}|}\frac{i\kappa_0 x_j y_j}{|\mathbf{x}|} + \cdots \equiv \text{constant} + U_j y_j + \cdots,$$

$$\text{where } U_j = \frac{e^{i\kappa_0|\mathbf{x}|}}{4\pi|\mathbf{x}|}\frac{i\kappa_0 x_j}{|\mathbf{x}|}, \qquad (3.4.4)$$

can be regarded as the velocity potential of a *uniform flow* at velocity U_j impinging on the solid.

At distances $|\mathbf{y}| \gg \ell$ from S, the distortion of this flow produced by the body must be small. Let it be represented by the velocity potential

$$\hat{G}'(\mathbf{y}, \mathbf{x}, \omega) = -U_j\varphi_j^*(\mathbf{y}) + \mathrm{O}(\kappa_0\ell)^2, \qquad \text{where } \varphi_j^*(\mathbf{y}) \to 0 \quad \text{when } |\mathbf{y}| \gg \ell.$$

The function φ_j^* has the dimensions of length and $\sim \ell$ in order of magnitude (Batchelor 1967). Then,

$$\hat{G}(\mathbf{y}, \mathbf{x}, \omega) = \hat{G}_0(\mathbf{y}, \mathbf{x}, \omega) + \hat{G}'(\mathbf{y}, \mathbf{x}, \omega) = \frac{-e^{i\kappa_0|\mathbf{x}|}}{4\pi|\mathbf{x}|} + U_j(y_j - \varphi_j^*(\mathbf{y})) + \cdots,$$

$$(3.4.5)$$

where the terms shown explicitly represent a potential flow past the body.

Near the body $\hat{G}(\mathbf{y}, \mathbf{x}, \omega)$ satisfies (3.4.2) with the right-hand side replaced by zero (because the source is in the far field). Hence,

$$U_j\nabla^2(y_j - \varphi_j^*(\mathbf{y})) + \mathrm{O}(\kappa_0\ell)^2 = 0.$$

But $U_j(y_j - \varphi_j^*(\mathbf{y})) = \mathrm{O}(\kappa_0\ell)$, and therefore, *correct to the neglect of small terms of order* $\mathrm{O}(\kappa_0\ell)^2$,

$$\nabla^2(y_j - \varphi_j^*(\mathbf{y})) = 0, \quad \text{i.e. } \nabla^2\varphi_j^*(\mathbf{y}) = 0,$$

where the rigid surface condition requires

$$\frac{\partial}{\partial y_n}(y_j - \varphi_j^*(\mathbf{y})) = 0 \quad \text{on } S. \tag{3.4.6}$$

Summarizing our conclusions from Equations (3.4.4)–(3.4.6): When $\mathbf{x}$ is in the acoustic far field, and $\mathbf{y}$ is close to the body

$$\hat{G}(\mathbf{x}, \mathbf{y}, \omega) = \frac{-e^{i\kappa_0|\mathbf{x}|}}{4\pi|\mathbf{x}|}\left(1 - \frac{i\kappa_0 x_j}{|\mathbf{x}|}(y_j - \varphi_j^*(\mathbf{y})) + O(\kappa_0\ell)^2\right),$$

$$\mathbf{y} \sim O(\ell), \quad |\mathbf{x}| \to \infty. \tag{3.4.7}$$

The first term in the large brackets represents the contribution from the spherical wave $\hat{G}_0(\mathbf{x}, \mathbf{y}, \omega)$ evaluated at $\mathbf{y} = \mathbf{0}$. The next term is $O(\kappa_0\ell)$ and includes a component $-i\kappa_0 x_j y_j/|\mathbf{x}|$ from the incident wave plus a correction $i\kappa_0 x_j \varphi_j^*(\mathbf{y})/|\mathbf{x}|$ produced by S.

The vector field

$$\mathbf{Y}(\mathbf{y}) \equiv \mathbf{y} - \boldsymbol{\varphi}^*(\mathbf{y}) \tag{3.4.8}$$

is called the **Kirchhoff vector** for the body; the jth component

$$Y_j(\mathbf{y}) \equiv y_j - \varphi_j^*(\mathbf{y})$$

satisfies Laplace's equation $\nabla^2 Y_j = 0$ with $\partial Y_j/\partial y_n = 0$ on S, and can be interpreted as the velocity potential of an incompressible flow past S that has unit speed in the j direction at large distances from S. The function $\varphi_j^*(\mathbf{y})$ decays with distance from S, and satisfies

$$\frac{\partial \varphi_j^*}{\partial y_n}(\mathbf{y}) = n_j \quad \text{on } S, \tag{3.4.9}$$

because $\partial y_j/\partial y_n \equiv n_i \partial y_j/\partial y_i = n_i \delta_{ij} = n_j$. Hence, $\varphi_j^*(\mathbf{y})$ is just the instantaneous velocity potential of the motion that would be produced by translational motion of S as a *rigid body* at unit speed in the j direction.

Definition

$$\hat{G}(\mathbf{x}, \mathbf{y}, \omega) = \frac{-e^{i\kappa_0|\mathbf{x}|}}{4\pi|\mathbf{x}|}\left(1 - \frac{i\kappa_0 x_j}{|\mathbf{x}|}(y_j - \varphi_j^*(\mathbf{y}))\right), \quad \mathbf{y} \sim O(\ell), \quad |\mathbf{x}| \to \infty,$$

$$\tag{3.4.10}$$

is called the compact Green's function for source points $\mathbf{y}$ near the body and observer positions $\mathbf{x}$ in the acoustic far field.

In Section 3.7, we shall introduce a very much more elegant representation of the compact Green's function that greatly expands its utility.

3.5 Compact Green's Function for a Rigid Sphere

Let the sphere have radius a and take the coordinate origin O at its center, as illustrated in Fig. 3.5.1. We have to determine the Kirchhoff vector whose jth component

$$Y_j(\mathbf{y}) = y_j - \varphi_j^*(\mathbf{y}) \quad \text{for } j = 1, 2, 3$$

is equal to the velocity potential of incompressible flow past the sphere having unit speed in the j direction at large distances from the sphere.

Consider the case $j = 1$ shown in the figure. The flow is evidently symmetric about the y_1 axis. Take spherical polar coordinates (r, ϑ, ϕ) with ϑ measured from the positive y_1 axis. Then, $y_1 = r \cos \vartheta$ and the condition (3.4.9) to be satisfied on the sphere is

$$\frac{\partial \varphi_1^*}{\partial r} = \cos \vartheta \quad \text{at } r = a. \tag{3.5.1}$$

The axisymmetry of the problem suggests that we look for a solution of Laplace's equation in the form

$$\varphi_1^* = \Psi(r) \cos \vartheta,$$

which satisfies the axisymmetric form of Laplace equation

$$\left(\frac{1}{r^2} \frac{\partial}{\partial r} \left(r^2 \frac{\partial}{\partial r} \right) + \frac{1}{r^2 \sin \vartheta} \frac{\partial}{\partial \vartheta} \left(\sin \vartheta \frac{\partial}{\partial \vartheta} \right) \right) \Psi(r) \cos \vartheta = 0,$$

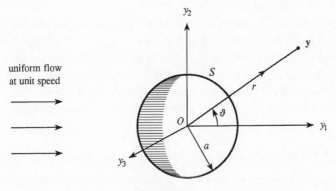

Fig. 3.5.1.

provided that

$$r^2 \frac{d^2 \Psi}{dr^2} + 2r \frac{d\Psi}{dr} - 2\Psi = 0.$$

The solutions of this equation are proportional to r^n where n is a root of the quadratic equation

$$n^2 + n - 2 = 0, \quad \text{i.e., } n = -2, 1.$$

Hence,

$$Y_1 \equiv y_1 - \varphi_1^* = r \cos \vartheta - \left(Ar + \frac{B}{r^2} \right) \cos \vartheta,$$

where A and B are constants. The condition that $\varphi_1^* \to 0$ as $r \to \infty$ implies that $A = 0$, and condition (3.5.1) supplies $B = -a^3/2$. Therefore,

$$Y_1 = r \cos \vartheta + \frac{a^3}{2r^2} \cos \vartheta \equiv y_1 \left(1 + \frac{a^3}{2r^3} \right).$$

Because of the symmetry of the sphere it is clear that we also have

$$Y_2 = y_2 \left(1 + \frac{a^3}{2r^3} \right), \quad Y_3 = y_3 \left(1 + \frac{a^3}{2r^3} \right), \quad r = |\mathbf{y}|.$$

Thus, the compact Green's function (3.4.10) for the sphere is

$$\hat{G}(\mathbf{x}, \mathbf{y}, \omega) = \frac{-e^{i\kappa_0 |\mathbf{x}|}}{4\pi |\mathbf{x}|} \left\{ 1 - \frac{i\kappa_0 x_j y_j}{|\mathbf{x}|} \left(1 + \frac{a^3}{2|\mathbf{y}|^3} \right) \right\}, \quad \mathbf{y} \sim O(a),$$

$$|\mathbf{x}| \to \infty. \quad (3.5.2)$$

This represents the far-field acoustic potential produced by a point source at $\mathbf{y}$ close to the sphere. Because $\kappa_0 |\mathbf{y}|$ is small the second term in the brace brackets is always *small* compared to 1. This appears to suggest that, after all, the sphere has a relatively small effect on the production of sound! This is certainly true for monopole sources, but most sources of interest in applications are dipoles or quadrupoles, and in these circumstances we shall see that it is the small, second term that determines the leading order approximation for the far-field sound.

3.5.1 Radiation from a Dipole Adjacent to a Compact Sphere

Let us apply the compact Green's function (3.5.2) to determine the far-field sound generated by a dipole source close to a sphere of radius $a \ll \lambda = $ acoustic

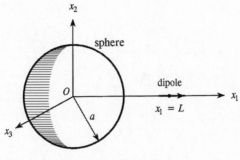

Fig. 3.5.2.

wavelength. With the origin at the center of the sphere, we consider the outgoing wave solution of

$$\left(\nabla^2 + \kappa_0^2\right)\hat{\varphi} = f_1 \frac{\partial}{\partial x_1}\{\delta(x_1 - L)\delta(x_2)\delta(x_3)\}, \quad \text{where } \frac{\partial\hat{\varphi}}{\partial x_n} = 0 \quad \text{on } |\mathbf{x}| = a.$$

The dipole is orientated in the x_1 direction and lies on the x_1 axis at $(L, 0, 0)$, as in Fig. 3.5.2. The solution is given by

$$\hat{\varphi}(\mathbf{x}, \omega) = \int f_1 \frac{\partial}{\partial y_1}\{\delta(y_1 - L)\delta(y_2)\delta(y_3)\}\,\hat{G}(\mathbf{x}, \mathbf{y}, \omega)\,d^3\mathbf{y},$$

where the integration is over the fluid, and $\partial\hat{G}/\partial x_n = 0$ on the sphere. The source term is zero everywhere except at $(L, 0, 0)$. To evaluate the integral we write

$$\hat{\varphi}(\mathbf{x}, \omega) = f_1 \int \frac{\partial}{\partial y_1}\{\hat{G}(\mathbf{x}, \mathbf{y}, \omega)\delta(y_1 - L)\delta(y_2)\delta(y_3)\}\,d^3\mathbf{y}$$

$$- f_1 \int \delta(y_1 - L)\delta(y_2)\delta(y_3)\frac{\partial\hat{G}}{\partial y_1}(\mathbf{x}, \mathbf{y}, \omega)\,d^3\mathbf{y}.$$

The first integral is zero because $\delta(y_1 - L) = 0$ on the boundaries of the region of integration, so that

$$\hat{\varphi}(\mathbf{x}, \omega) = -f_1 \left(\frac{\partial\hat{G}}{\partial y_1}(\mathbf{x}, \mathbf{y}, \omega)\right)_{\mathbf{y}=(L,0,0)}. \qquad (3.5.3)$$

Thus far the calculation is exact. To determine the solution in the far field given that the sphere is acoustically compact we use the compact approximation (3.5.2) for $\hat{G}(\mathbf{x}, \mathbf{y}, \omega)$. We see immediately that the differentiation with respect to y_1 will be applied only to the small second term in the braces of (3.5.2), giving

$$\hat{\varphi}(\mathbf{x}, \omega) = \frac{-i\kappa_0 f_1 x_j e^{i\kappa_0|\mathbf{x}|}}{4\pi |\mathbf{x}|^2} \left\{ \frac{\partial}{\partial y_1} \left[y_j \left(1 + \frac{a^3}{2|\mathbf{y}|^3} \right) \right] \right\}_{\mathbf{y}=(L,0,0)}$$

$$= \frac{-i\kappa_0 f_1 x_1 e^{i\kappa_0|\mathbf{x}|}}{4\pi |\mathbf{x}|^2} \left(1 - \frac{a^3}{L^3} \right)$$

$$= \frac{-i\kappa_0 f_1 \cos\theta\, e^{i\kappa_0|\mathbf{x}|}}{4\pi |\mathbf{x}|} \left(1 - \frac{a^3}{L^3} \right), \quad |\mathbf{x}| \to \infty, \ \kappa_0 a \ll 1,$$

where θ is the angle between the x_1 axis and the $\mathbf{x}$ direction (so that $x_1 = |\mathbf{x}| \cos\theta$).

By setting $a = 0$ in this formula, we recover the far field (3.2.10) of a dipole source in the absence of the sphere. The presence of the sphere accordingly *reduces* the amplitude of the sound relative to that produced by a free-field dipole. The amplitude is zero when $L \to a$, because in this limit the surface of the sphere is effectively plane in the vicinity of the dipole and an equal and opposite image dipole is formed in the sphere. The net radiation is therefore equivalent to that produced by a *quadrupole source*, and to calculate the sound in this case it would be necessary to use a more accurate approximation to $\hat{G}(\mathbf{x}, \mathbf{y}, \omega)$. This conclusion applies only to dipoles oriented *radially* with respect to the sphere (see Problem 1), but it is also true for *any* compact rigid surface when a dipole orientated in the direction of the local surface normal approaches the surface.

3.5.2 Sound Produced by a Vibrating Sphere

Let the surface S of a fixed body execute small amplitude vibrations with normal velocity $v_n(\mathbf{x}, \omega)$ (Fig. 3.5.3). The fluid motion is the same as that generated by a distribution of monopoles of strength $v_n(\mathbf{x}, \omega)$ per unit area of S when S is

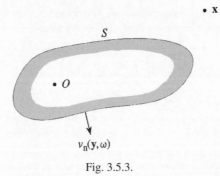

Fig. 3.5.3.

assumed to be stationary (*rigid*). The corresponding source strength $\hat{q}(\mathbf{x}, \omega)$ in the inhomogeneous Helmholtz equation (3.2.1) is

$$\hat{q}(\mathbf{x}, \omega) = v_\text{n}(\mathbf{x}, \omega)\delta(s_\perp - \epsilon), \quad (\epsilon \to +0),$$

where $s_\perp$ is distance measured in the normal direction from S into the fluid, and $\epsilon > 0$ places the sources just within the fluid adjacent to S. The velocity potential $\hat{\varphi}(\mathbf{x}, \omega)$ is therefore

$$\hat{\varphi}(\mathbf{x}, \omega) = \int_\text{fluid} v_\text{n}(\mathbf{y}, \omega)\delta(s_\perp - \epsilon)\hat{G}(\mathbf{x}, \mathbf{y}, \omega)\, d^3\mathbf{y} \quad (\epsilon \to +0)$$

$$= \oint_S v_\text{n}(\mathbf{y}, \omega)\hat{G}(\mathbf{x}, \mathbf{y}, \omega)\, dS(\mathbf{y}) \quad \text{where } \frac{\partial \hat{G}}{\partial x_\text{n}}(\mathbf{x}, \mathbf{y}, \omega) = 0 \quad \text{on } S.$$

$$(3.5.4)$$

Consider the sound produced when the sphere of Fig. 3.5.1 vibrates with small amplitude about its undisturbed position centred at the origin with velocity $\hat{U}(\omega)e^{-i\omega t}$ along the x_1 axis. Then,

$$v_\text{n}(\mathbf{y}, \omega) = \hat{U}(\omega)\cos\vartheta.$$

If the vibrations are at sufficiently low frequency, the sphere will be compact, and when the observer at $\mathbf{x}$ is in the acoustic far field the integral in (3.5.4) can be evaluated using the compact approximation (3.5.2) for $\hat{G}(\mathbf{x}, \mathbf{y}, \omega)$:

$$\hat{\varphi}(\mathbf{x}, \omega)$$
$$\approx \frac{-e^{i\kappa_0|\mathbf{x}|}}{4\pi|\mathbf{x}|}\left[\oint_S v_\text{n}(\mathbf{y}, \omega)\, dS(\mathbf{y}) - \frac{i\kappa_0 x_j}{|\mathbf{x}|}\oint_S y_j\left(1 + \frac{a^3}{2|\mathbf{y}|^3}\right)v_\text{n}(\mathbf{y}, \omega)\, dS(\mathbf{y})\right].$$

The first integral represents the net volume flux through S and vanishes identically for rigid body translational motion. The second integral is nonzero only for $j = 1$, when $y_1 = a\cos\vartheta$ and $|\mathbf{y}| = a$ on S, and we can take $dS = 2\pi a^2 \sin\vartheta\, d\vartheta$ (so that the surface integral becomes $3\pi a^3 \hat{U}(\omega)\int_0^\pi \cos^2\vartheta \sin\vartheta\, d\vartheta = 2\pi a^3 \hat{U}(\omega)$). Hence,

$$\hat{\varphi}(\mathbf{x}, \omega) \approx \frac{i\kappa_0\hat{U}(\omega)a^3 x_1 e^{i\kappa_0|\mathbf{x}|}}{2|\mathbf{x}|^2} \equiv \frac{i\omega\hat{U}(\omega)a^3\cos\theta e^{i\kappa_0|\mathbf{x}|}}{2c_0|\mathbf{x}|}, \quad |\mathbf{x}| \to \infty,$$

where θ is the angle between the x_1 axis and the radiation direction $\mathbf{x}$ (see Fig. 1.7.1).

The solution for a sphere oscillating at an arbitrary time dependent velocity $U(t)$ can be derived from this result provided the sphere remains compact. This means that if we write

$$U(t) = \int_{-\infty}^{\infty} \hat{U}(\omega)e^{-i\omega t}\, d\omega,$$

then $\hat{U}(\omega) \neq 0$ only for $\kappa_0 a \ll 1$. If this condition is satisfied we can use the Formulae (3.2.2) and (3.2.3) to obtain the time-dependent velocity potential in the form

$$\begin{aligned}
\varphi(\mathbf{x}, t) &\approx \frac{a^3 \cos\theta}{2c_0|\mathbf{x}|} \int_{-\infty}^{\infty} i\omega\hat{U}(\omega)e^{-i\omega(t-|\mathbf{x}|/c_0)}\, d\omega \\
&= \frac{-a^3 \cos\theta}{2c_0|\mathbf{x}|} \frac{\partial}{\partial t} \int_{-\infty}^{\infty} \hat{U}(\omega)e^{-i\omega(t-|\mathbf{x}|/c_0)}\, d\omega \\
&= \frac{-a^3 \cos\theta}{2c_0|\mathbf{x}|} \frac{\partial U}{\partial t}(t - |\mathbf{x}|/c_0), \quad |\mathbf{x}| \to \infty.
\end{aligned}$$

This agrees with the far-field result obtained in Section 1.7, and confirms the model used there in which the vibrating sphere was replaced by a point dipole of strength $2\pi a^3 U(t)$ at its center.

3.6 Compact Green's Function for Cylindrical Bodies

The reciprocal calculation of the Green's function described in Section 3.4 for the compact body in Fig. 3.4.1 can be immediately extended to the case of a cylindrical body of *compact cross section*.

Figure 3.6.1 illustrates the situation for an infinite circular cylinder of radius a whose axis lies along the y_3 axis, and whose diameter $2a \sim \ell$ is acoustically compact. The source point $\mathbf{y}$ is adjacent to the cylinder and for the moment (see Section 3.7) is assumed to be within an axial distance $|y_3| \ll \lambda$ from the coordinate origin O. In this region the Expansion (3.4.7) remains valid with $\varphi_3^*(\mathbf{y}) \equiv 0$, because the impinging flow described by the velocity potential (3.4.4) for $j = 3$ is unaffected by the cylinder. Hence, we can take

$$\hat{G}(\mathbf{x}, \mathbf{y}, \omega) = \frac{-e^{i\kappa_0|\mathbf{x}|}}{4\pi|\mathbf{x}|} \left(1 - \frac{i\kappa_0 x_j Y_j}{|\mathbf{x}|}\right), \quad \mathbf{y} \sim O(\ell), \quad |\mathbf{x}| \to \infty, \quad (3.6.1)$$

where the Kirchhoff vector $\mathbf{Y}$ has the components

$$Y_1 = y_1 - \varphi_1^*(\mathbf{y}), \qquad Y_2 = y_2 - \varphi_2^*(\mathbf{y}), \qquad Y_3 = y_3. \quad (3.6.2)$$

3.6.1 Circular Cylinder

The potentials $\varphi_1^*(\mathbf{y})$, $\varphi_2^*(\mathbf{y})$ for the circular cylinder of radius a can be found by the method of Section 3.5.

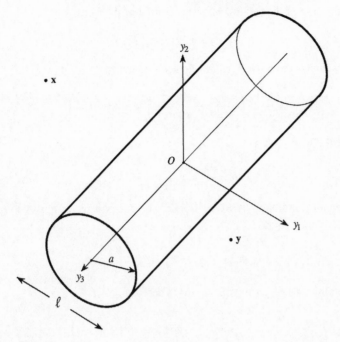

Fig. 3.6.1.

For $j = 1$ the flow is symmetric about the y_1 axis and is independent of the spanwise coordinate y_3 (Fig. 3.6.2). Using polar coordinates $(y_1, y_2) = r(\cos \vartheta, \sin \vartheta)$, the condition (3.4.9) to be satisfied on the cylinder is

$$\frac{\partial \varphi_1^*}{\partial r} = \cos \vartheta \quad \text{at } r = a. \tag{3.6.3}$$

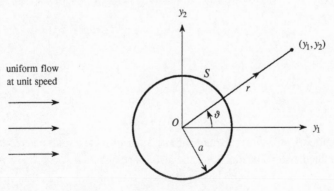

Fig. 3.6.2.

As in the case of the sphere, we try a solution of the form

$$\varphi_1^* = \Psi(r)\cos\vartheta,$$

which satisfies the polar form of Laplace's equation

$$\left(\frac{1}{r}\frac{\partial}{\partial r}\left(r\frac{\partial}{\partial r}\right) + \frac{1}{r^2}\frac{\partial^2}{\partial\vartheta^2}\right)\Psi(r)\cos\vartheta = 0,$$

provided that

$$r^2\frac{d^2\Psi}{dr^2} + r\frac{d\Psi}{dr} - \Psi = 0.$$

The general solution is $\Psi = Ar + B/r$. The component Ar must be rejected because it does not decay as $r \to \infty$. Therefore,

$$Y_1 \equiv y_1 - \varphi_1^* = r\cos\vartheta - \frac{B}{r}\cos\vartheta,$$

and condition (3.6.3) yields $B = -a^2$. Therefore,

$$Y_1 = r\cos\vartheta + \frac{a^2}{r}\cos\vartheta \equiv y_1\left(1 + \frac{a^2}{r^2}\right).$$

Similarly,

$$Y_2 = y_2\left(1 + \frac{a^2}{r^2}\right).$$

Hence, the **compact Green's function for a circular cylinder,** with source near the origin, is

$$\hat{G}(\mathbf{x}, \mathbf{y}, \omega) = \frac{-e^{i\kappa_0|\mathbf{x}|}}{4\pi|\mathbf{x}|}\left(1 - \frac{i\kappa_0 x_j Y_j}{|\mathbf{x}|}\right), \quad \mathbf{y} \sim O(\ell), \quad |\mathbf{x}| \to \infty, \quad (3.6.4)$$

where

$$Y_j = y_j\left(1 + \frac{a^2}{y_1^2 + y_2^2}\right), \quad j = 1, 2; \quad Y_3 = y_3. \quad (3.6.5)$$

3.6.2 Rigid Strip

The rigid strip of chord $2a$ and infinite span provides a simple model of a sharp-edged airfoil. In Fig. 3.6.3 the airfoil occupies $-a < y_1 < a$, $y_2 = 0$, $-\infty < y_3 < \infty$. The airfoil has no influence on a uniform mean flow in the y_1-direction, nor on one in the y_3-direction, so that potential functions $\varphi_1^*(\mathbf{y}) \equiv 0$ and $\varphi_3^*(\mathbf{y}) \equiv 0$.

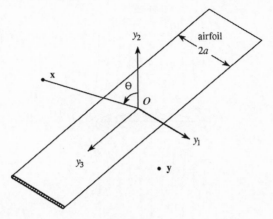

Fig. 3.6.3.

The potential $\varphi_2^*(\mathbf{y}) \equiv \varphi_2^*(y_1, y_2)$ can be determined by the method of conformal transformation. (Readers unfamiliar with this procedure should consult Section 4.5.) If $z = y_1 + iy_2$, the cross section of the airfoil in the z plane is mapped onto the circular cylinder $|Z| = a$ in the Z plane by the transformation

$$Z = z + \sqrt{z^2 - a^2}.$$

Because $Z \sim 2z$ as $|z| \to \infty$ a uniform flow at unit speed in the y_2 direction in the z plane at large distances from the airfoil corresponds to a uniform flow at speed $\frac{1}{2}$ in the direction of the imaginary Z axis at large distances from the cylinder. This flow can be found by the method discussed above for the circular cylinder (or see Example 3 of Section 4.5), and determines $Y_2 = y_2 - \varphi_2^*(y_1, y_2) = \mathrm{Re}[w(z)]$, where w is the complex potential

$$w(z) = -\frac{i}{2}\left(Z - \frac{a^2}{Z}\right)$$

$$= -\frac{i}{2}\left(z + \sqrt{z^2 - a^2} - \frac{a^2}{z + \sqrt{z^2 - a^2}}\right)$$

$$= -i\sqrt{z^2 - a^2}.$$

Thus, the **compact Green's function for a strip,** with source near the origin, is given by

$$\hat{G}(\mathbf{x}, \mathbf{y}, \omega) = \frac{-e^{i\kappa_0 |\mathbf{x}|}}{4\pi |\mathbf{x}|}\left(1 - \frac{i\kappa_0 x_j Y_j}{|\mathbf{x}|}\right), \quad \mathbf{y} \sim \mathrm{O}(a), \quad |\mathbf{x}| \to \infty, \quad (3.6.6)$$

where the components of the Kirchhoff vector are

$$Y_1 = y_1, \quad Y_2 = \mathrm{Re}(-i\sqrt{z^2 - a^2}), \quad Y_3 = y_3, \quad z = y_1 + iy_2. \quad (3.6.7)$$

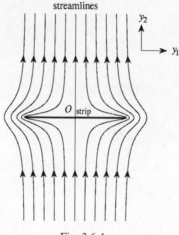

streamlines

Fig. 3.6.4.

Figure 3.6.4 depicts the streamline pattern of the flow past the strip defined by the velocity potential $Y_2(\mathbf{y})$. The streamlines crowd together and change very rapidly near the sharp edges. This is an indication that edges can be important sources of noise when located in the near field of a dipole or quadrupole (or any higher order multipole) source, because ∇Y_2 becomes very large there.

Example Calculate the far-field velocity potential when

$$\left(\nabla^2 + \kappa_0^2\right)\hat{\varphi} = f_2 \frac{\partial}{\partial x_2}\left[\delta(x_1 - L)\delta(x_2)\delta(x_3)\right], \quad L > a, \quad \kappa_0 L \ll 1,$$

where $\dfrac{\partial \hat{\varphi}}{\partial x_2} = 0$ on the airfoil $-a < x_1 < a, \quad x_2 = 0, \quad -\infty < x_3 < \infty.$

The dipole source is orientated in the x_2 direction and is positioned just to the right of the edge at $y_1 = a$ in Fig. 3.6.3. The solution is given by the following form of Equation (3.5.3)

$$\hat{\varphi}(\mathbf{x}, \omega) = -f_2 \left(\frac{\partial \hat{G}}{\partial y_2}(\mathbf{x}, \mathbf{y}, \omega)\right)_{\mathbf{y}=(L,0,0)} \approx -\frac{if_2\kappa_0 x_2 e^{i\kappa_0|\mathbf{x}|}}{4\pi|\mathbf{x}|^2}\left(\frac{\partial Y_2}{\partial y_2}\right)_{\mathbf{y}=(L,0,0)}$$

where, from (3.6.7),

$$\frac{\partial Y_2}{\partial y_2} = \mathrm{Re}\left(-i\frac{\partial}{\partial y_2}\sqrt{z^2 - a^2}\right), \quad z = y_1 + iy_2$$

$$= \mathrm{Re}\left(\frac{z}{\sqrt{z^2 - a^2}}\right),$$

Therefore,

$$\hat{\varphi}(\mathbf{x}, \omega) \approx -\frac{i f_2 \kappa_0 x_2 e^{i\kappa_0|\mathbf{x}|}}{4\pi |\mathbf{x}|^2} \frac{L}{\sqrt{L^2 - a^2}}$$

$$= -\frac{i f_2 \kappa_0 L \cos \Theta \, e^{i\kappa_0|\mathbf{x}|}}{4\pi |\mathbf{x}| \sqrt{L^2 - a^2}}, \quad |\mathbf{x}| \to \infty,$$

where $\Theta = \cos^{-1}(x_2/|\mathbf{x}|)$ is the angle between the normal to the strip and the radiation direction $(\mathbf{x}/|\mathbf{x}|)$ indicated in Fig. 3.6.3.

The amplitude of the sound is increased by a factor $L/\sqrt{L^2 - a^2}$ relative to that produced by the same dipole in free space, and is unbounded as $L \to a$, when the dipole approaches the edge.

3.7 Symmetric Compact Green's Function

The definition (3.4.10) of the compact Green's function can be recast to exhibit the reciprocal nature of the source and observer positions $\mathbf{y}$ and $\mathbf{x}$. To do this we first observe that, for a body of characteristic diameter ℓ,

$$Y_j(\mathbf{y}) = y_j - \varphi_j^*(\mathbf{y}) \sim O(\ell),$$

and, therefore, that $\kappa_0 Y_j \ll 1$. Hence,

$$\hat{G}(\mathbf{x}, \mathbf{y}, \omega) \approx \frac{-e^{i\kappa_0|\mathbf{x}|}}{4\pi |\mathbf{x}|} \left(1 - \frac{i\kappa_0 x_j Y_j}{|\mathbf{x}|}\right) = \frac{-e^{i\kappa_0|\mathbf{x}|}}{4\pi |\mathbf{x}|} \left(1 - \frac{i\kappa_0 \mathbf{x} \cdot \mathbf{Y}}{|\mathbf{x}|}\right)$$

$$\approx \frac{-1}{4\pi |\mathbf{x}|} e^{i\kappa_0|\mathbf{x}| - \frac{i\kappa_0 \mathbf{x} \cdot \mathbf{Y}}{|\mathbf{x}|}},$$

$$\approx \frac{-e^{i\kappa_0|\mathbf{x} - \mathbf{Y}|}}{4\pi |\mathbf{x}|}, \quad \mathbf{Y} \sim O(\ell), \quad |\mathbf{x}| \to \infty, \qquad (3.7.1)$$

where on the last line we have used the usual far-field approximation (1.9.2)

$$|\mathbf{x} - \mathbf{Y}| \approx |\mathbf{x}| - \frac{\mathbf{x} \cdot \mathbf{Y}}{|\mathbf{x}|}, \quad |\mathbf{x}| \to \infty.$$

Now let $\mathbf{X}(\mathbf{x})$ denote the Kirchhoff vector for the body expressed in terms of $\mathbf{x}$, i.e., let

$$X_j(\mathbf{x}) = x_j - \varphi_j^*(\mathbf{x}). \qquad (3.7.2)$$

Then, because $\varphi_j^*(\mathbf{x}) \to 0$ as $|\mathbf{x}| \to \infty$, we also have $|\mathbf{X}| \sim |\mathbf{x}|$ as $|\mathbf{x}| \to \infty$, and, therefore, from (1.9.2) and (1.9.3)

$$
\left.
\begin{aligned}
|\mathbf{X} - \mathbf{Y}| &\approx |\mathbf{x}| - \frac{\mathbf{x} \cdot \mathbf{Y}}{|\mathbf{x}|} \\[2mm]
\frac{1}{|\mathbf{X} - \mathbf{Y}|} &\approx \frac{1}{|\mathbf{x}|} + \frac{\mathbf{x} \cdot \mathbf{Y}}{|\mathbf{x}|^3} \approx \frac{1}{|\mathbf{x}|}
\end{aligned}
\right\} \quad \text{when } |\mathbf{x}| \to \infty. \quad (3.7.3)
$$

Thus, to the same approximation, (3.7.1) can be written

$$
\hat{G}(\mathbf{x}, \mathbf{y}, \omega) \approx \frac{-e^{i\kappa_0 |\mathbf{X} - \mathbf{Y}|}}{4\pi |\mathbf{X} - \mathbf{Y}|}, \quad \mathbf{Y} \sim \mathrm{O}(\ell), \quad |\mathbf{x}| \to \infty.
$$

This result is the basis of our revised definition of the

Compact Green's Function for the Inhomogeneous Helmholtz Equation

$$
\hat{G}(\mathbf{x}, \mathbf{y}, \omega) = \frac{-e^{i\kappa_0 |\mathbf{X} - \mathbf{Y}|}}{4\pi |\mathbf{X} - \mathbf{Y}|}, \quad (3.7.4)
$$

where $\mathbf{X} = \mathbf{x} - \boldsymbol{\varphi}^*(\mathbf{x})$, $\mathbf{Y} = \mathbf{y} - \boldsymbol{\varphi}^*(\mathbf{y})$ are the Kirchhoff vectors for the body expressed respectively in terms of $\mathbf{x}$ and $\mathbf{y}$. The components X_j and Y_j are the velocity potential of incompressible flow past the body having unit speed in the j direction at large distances from the body; φ_j^* is the velocity potential of the incompressible flow that would be produced by rigid body motion of S at unit speed in the j direction.

Our generalized definition clearly satisfies the reciprocal theorem. Also, because of the symmetrical way in which $\mathbf{x}$ and $\mathbf{y}$ enter this formula we may now remove any restriction on the position of the coordinate origin. The approximation is valid for arbitrary source and observer locations provided that *at least one of them* lies in the far field of the body. When *both* $\mathbf{x}$ and $\mathbf{y}$ are in the far field (so that $\mathbf{X} \sim \mathbf{x}$ and $\mathbf{Y} \sim \mathbf{y}$) predictions made with the compact Green's function will be the same as when the body is *absent*. This is because for distant sources the amplitude of the sound scattered by a compact rigid object is $\mathrm{O}((\kappa_0\ell)^2)$ smaller than the incident sound, that is, is of *quadrupole* intensity (Lighthill 1978; Howe 1998a). When $\mathbf{x}$ is close to the body the source must be in the far field; $\hat{G}(\mathbf{x}, \mathbf{y}, \omega)$ then determines the modification by the body of low frequency sound received by an observer near the body.

The definition (3.7.4) is easily recalled because it is an obvious generalization of the free space Green's function (3.2.6). In applications it is necessary to remember also that it is valid for determining only the leading order approximation to the surface monopole and dipole sources induced on the body by neighboring sources in the fluid. In practice this means that when used in calculations $\hat{G}(\mathbf{x}, \mathbf{y}, \omega)$ will normally be expanded **only to first order** in the Kirchhoff source vector $\mathbf{Y}(\mathbf{y})$.

We can go further and use the formula (3.1.6) relating Green's functions for the wave equation and the Helmholtz equation to derive the compact approximation for Green's function of the wave equation:

$$
\begin{aligned}
G(\mathbf{x}, \mathbf{y}, t - \tau) &= \frac{-1}{2\pi} \int_{-\infty}^{\infty} \hat{G}(\mathbf{x}, \mathbf{y}, \omega) e^{-i\omega(t-\tau)} \, d\omega \\
&\approx \frac{1}{8\pi^2 |\mathbf{X} - \mathbf{Y}|} \int_{-\infty}^{\infty} e^{-i\omega\left(t-\tau-\frac{|\mathbf{X}-\mathbf{Y}|}{c_0}\right)} d\omega \\
&= \frac{1}{4\pi |\mathbf{X} - \mathbf{Y}|} \delta\left(t - \tau - \frac{|\mathbf{X} - \mathbf{Y}|}{c_0}\right).
\end{aligned}
$$

This remarkable result is formally identical with the classical free space Green's function (1.6.2) with $\mathbf{X}$, $\mathbf{Y}$ substituted for $\mathbf{x}$, $\mathbf{y}$. However, its use is subject to the same restrictions as (3.7.4), and it will be valid only when applied to time-dependent source terms producing sound whose wavelength is large compared to the characteristic body dimension ℓ. With this understanding we can define the

Compact Green's Function for the Wave Equation

$$
G(\mathbf{x}, \mathbf{y}, t - \tau) = \frac{1}{4\pi |\mathbf{X} - \mathbf{Y}|} \delta\left(t - \tau - \frac{|\mathbf{X} - \mathbf{Y}|}{c_0}\right), \tag{3.7.5}
$$

where $\mathbf{X} = \mathbf{x} - \boldsymbol{\varphi}^*(\mathbf{x})$, $\mathbf{Y} = \mathbf{y} - \boldsymbol{\varphi}^*(\mathbf{y})$ are Kirchhoff vectors for the body. The components X_j and Y_j are the velocity potential of incompressible flow past the body having unit speed in the j direction at large distances from the body; φ_j^* is the velocity potential of the incompressible flow that would be produced by rigid body motion of S at unit speed in the j direction.

3.8 Low-Frequency Radiation from a Vibrating Body

The compact Green's function (3.7.5) for the wave equation will now be used to give a complete theory of the low-frequency sound produced by a vibrating body. The maximum frequency of the vibrations must be small enough to ensure that the body (or its cross section, in the case of vibrating a cylinder) is acoustically compact. The argument follows closely the discussion of the vibrating sphere in Section 3.5, except that we now work directly with time dependent quantities.

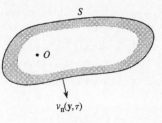

Fig. 3.8.1.

Let the closed surface S (Fig. 3.8.1) vibrate with normal velocity $v_n(\mathbf{x}, t)$. As before, the velocity potential in the fluid (governed by Equation (3.1.1)) is the same as that generated by the distribution of volume sources

$$q(\mathbf{x}, t) = v_n(\mathbf{x}, t)\delta(s_\perp - \epsilon) \quad (\epsilon \to +0) \tag{3.8.1}$$

distributed over S regarded as a *rigid, stationary* surface, where $s_\perp$ is distance measured in the normal direction from S *into* the fluid. The velocity potential $\varphi(\mathbf{x}, t)$ is given exactly by

$$\varphi(\mathbf{x}, t) = -\int_{-\infty}^{\infty} \int_{\text{fluid}} v_n(\mathbf{y}, \tau)\delta(s_\perp - \epsilon)G(\mathbf{x}, \mathbf{y}, t - \tau)\,d^3\mathbf{y}\,d\tau \quad (\epsilon \to +0)$$

$$= -\int_{-\infty}^{\infty} \oint_S v_n(\mathbf{y}, \tau)G(\mathbf{x}, \mathbf{y}, t - \tau)\,dS(\mathbf{y})\,d\tau,$$

$$\text{where} \quad \frac{\partial G}{\partial x_n}(\mathbf{x}, \mathbf{y}, t - \tau) = 0 \quad \text{on } S. \tag{3.8.2}$$

At low frequencies the first approximation to the far-field sound is obtained by replacing $G(\mathbf{x}, \mathbf{y}, t - \tau)$ in (3.8.2) by its compact approximation (3.7.5). The details are given below; they illustrate the general procedure that should be adopted when using the compact Green's function (in particular, the technique of expanding to first order in $\mathbf{Y}$):

$$\varphi(\mathbf{x}, t) \approx -\int_{-\infty}^{\infty} \int_{\text{fluid}} \frac{v_n(\mathbf{y}, \tau)\delta(s_\perp - \epsilon)}{4\pi|\mathbf{X} - \mathbf{Y}|}\delta\left(t - \tau - \frac{|\mathbf{X} - \mathbf{Y}|}{c_0}\right)d^3\mathbf{y}\,d\tau$$

$$(\epsilon \to +0), \quad |\mathbf{x}| \to \infty$$

$$= -\int_{-\infty}^{\infty} \oint_S \frac{v_n(\mathbf{y}, \tau)}{4\pi|\mathbf{X} - \mathbf{Y}|}\delta\left(t - \tau - \frac{|\mathbf{X} - \mathbf{Y}|}{c_0}\right)dS(\mathbf{y})\,d\tau,$$

$$= -\frac{1}{4\pi|\mathbf{x}|}\int_{-\infty}^{\infty} \oint_S v_n(\mathbf{y}, \tau)\delta\left(t - \tau - \frac{|\mathbf{x}|}{c_0} + \frac{\mathbf{x} \cdot \mathbf{Y}}{c_0|\mathbf{x}|}\right)dS(\mathbf{y})\,d\tau$$

$$(\mathbf{X} \sim \mathbf{x} \text{ as } |\mathbf{x}| \to \infty)$$

$$= -\frac{1}{4\pi |\mathbf{x}|} \int_{-\infty}^{\infty} \oint_S v_n(\mathbf{y}, \tau) \left\{ \delta \left(t - \tau - \frac{|\mathbf{x}|}{c_0} \right) \right.$$

$$\left. + \delta' \left(t - \tau - \frac{|\mathbf{x}|}{c_0} \right) \frac{x_j Y_j}{c_0 |\mathbf{x}|} \right\} dS(\mathbf{y}) \, d\tau,$$

where the prime denotes differentiation with respect to time. Performing the integration with respect to τ:

$$\varphi(\mathbf{x}, t) \approx -\frac{1}{4\pi |\mathbf{x}|} \oint_S v_n \left(\mathbf{y}, t - \frac{|\mathbf{x}|}{c_0} \right) dS(\mathbf{y})$$

$$-\frac{x_j}{4\pi c_0 |\mathbf{x}|^2} \frac{\partial}{\partial t} \oint_S v_n \left(\mathbf{y}, t - \frac{|\mathbf{x}|}{c_0} \right) Y_j(\mathbf{y}) \, dS(\mathbf{y}).$$

The first integral represents an omnidirectional *monopole* sound wave, and is nonzero only if the volume enclosed by S changes with time (i.e., only for a *pulsating* body). It is then the most important component of the far field sound – the second integral is smaller by a factor $\sim O(\omega \ell / c_0) \ll 1$ (because $\partial / \partial t \sim \omega$ and $Y_j \sim \ell$).

The monopole term vanishes for a *rigid body* executing small amplitude translational oscillations at velocity $\mathbf{U}(t)$, say. Then,

$$v_n(\mathbf{y}, \tau) = \mathbf{n}(\mathbf{y}) \cdot \mathbf{U}(\tau) = n_i(\mathbf{y}) U_i(\tau),$$

where $\mathbf{n}(\mathbf{y})$ is the surface normal directed into the fluid. Making the substitution $Y_j = y_j - \varphi_j^*(\mathbf{y})$ in the second integral we obtain an acoustic field of *dipole* type, given by

$$\varphi(\mathbf{x}, t) \approx -\frac{x_j}{4\pi c_0 |\mathbf{x}|^2} \frac{\partial U_i}{\partial t} \left(t - \frac{|\mathbf{x}|}{c_0} \right) \oint_S n_i(\mathbf{y}) Y_j(\mathbf{y}) \, dS(\mathbf{y}) \tag{3.8.3}$$

$$= -\frac{x_j}{4\pi c_0 |\mathbf{x}|^2} \frac{\partial U_i}{\partial t} \left(t - \frac{|\mathbf{x}|}{c_0} \right) \oint_S [n_i y_j - n_i \varphi_j^*] \, dS, \quad |\mathbf{x}| \to \infty. \tag{3.8.4}$$

Example: The vibrating sphere Consider a rigid sphere of radius a centred at the origin and oscillating in the x_1 direction at velocity $U(t)$ (Fig. 3.8.2). Then $\mathbf{U} = (U, 0, 0)$, and it is only necessary to take $i = 1$ in (3.8.3) or (3.8.4).

In terms of spherical polar coordinates (r, ϑ, ϕ) we have

$$\mathbf{y} = r(\cos \vartheta, \sin \vartheta \cos \phi, \sin \vartheta \sin \phi)$$

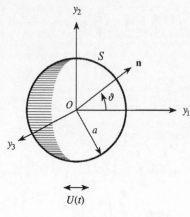

Fig. 3.8.2.

Therefore,

$$\mathbf{Y} \equiv \mathbf{y}\left(1 + \frac{a^3}{2|\mathbf{y}|^3}\right) = \frac{3a}{2}(\cos\vartheta, \sin\vartheta\cos\phi, \sin\vartheta\sin\phi)$$

on the sphere and $n_1 = \cos\vartheta$

Hence,

$$\oint_S n_1 Y_j \, dS = \frac{3a^3}{2} \oint_S (\cos\vartheta, \sin\vartheta\cos\phi, \sin\vartheta\sin\phi)\cos\vartheta \sin\vartheta \, d\vartheta \, d\phi$$

$$= \begin{cases} 2\pi a^3, & j = 1, \\ 0, & j = 2, 3 \end{cases}$$

and, therefore, (3.8.3) becomes

$$\varphi(\mathbf{x}, t) \approx \frac{-a^3\cos\theta}{2c_0|\mathbf{x}|}\frac{\partial U}{\partial t}(t - |\mathbf{x}|/c_0), \quad |\mathbf{x}| \to \infty, \quad \text{and} \quad \frac{x_1}{|\mathbf{x}|} = \cos\theta,$$

$$(3.8.5)$$

which is the result already obtained in Section 3.5 using the solution derived from the Helmholtz equation.

3.8.1 Far Field Pressure Produced by a Vibrating Body

A more general and illuminating discussion of the low-frequency sound produced by a vibrating rigid body can be given in terms of the *added mass tensor* M_{ij} (Batchelor 1967), defined by the surface integral

$$M_{ij} = -\rho_0 \oint_S n_i \varphi_j^* \, dS.$$

The Condition (3.4.9) satisfied by φ_j^* on S implies that $M_{ij} = M_{ji}$, because n_i can be replaced in the integrand by $\partial \varphi_i^* / \partial y_n$, and

$$M_{ij} = -\rho_0 \oint_S n_i \varphi_j^* \, dS = -\rho_0 \oint_S \frac{\partial \varphi_i^*}{\partial y_n} \varphi_j^* \, dS \equiv -\rho_0 \oint_S \varphi_i^* \frac{\partial \varphi_j^*}{\partial y_n} dS = M_{ji}.$$

(3.8.6)

The final integral is deduced from the second by referring to Fig. 3.3.1, recalling that $\nabla^2 \varphi_j^* = \nabla^2 \varphi_i^* = 0$, and applying the divergence theorem as follows:

$$\oint_{S+\Sigma} \left(\frac{\partial \varphi_i^*}{\partial y_n} \varphi_j^* - \varphi_i^* \frac{\partial \varphi_j^*}{\partial y_n} \right) dS = \int_{\text{fluid}} (\varphi_i^* \nabla^2 \varphi_j^* - \varphi_j^* \nabla^2 \varphi_i^*) \, d^3 \mathbf{y} \equiv 0.$$

The integration over Σ vanishes as the surface recedes to infinity (Batchelor 1967) because

$$\varphi_{i,j}^*(\mathbf{y}) \sim O\left(\frac{1}{|\mathbf{y}|^2} \right) \quad \text{as } |\mathbf{y}| \to \infty.$$

By evaluating the net force on S produced by the unsteady surface pressure (or by the method described below in Section 4.4) it can be verified that when the body translates at velocity $\mathbf{U}(t)$ without rotation in an ideal, *incompressible* fluid, it exerts a force on the fluid in the i direction given by

$$F_i(t) = M_{ij} \frac{dU_j}{dt}.$$

(3.8.7)

For a body of mass m, this means that when an external force $\mathcal{F}_i$ acts through its centre of mass, the equation of motion of the body can be written

$$(m\delta_{ij} + M_{ij}) \frac{dU_j}{dt} = \mathcal{F}_i.$$

The added mass tensor determines the effective mass of fluid dragged along by the body in its accelerated motion. The inertia of this fluid, in addition to that of the body, must also be overcome by the force $\mathcal{F}$ when the body accelerates. In general, however, a *couple* must also be applied to the translating body to counter a rotational torque also exerted on the body by the fluid (see Batchelor (1967) for further discussion).

Let us now apply these concepts to determine from (3.8.4) the sound pressure produced by a rigid compact body executing small amplitude translational oscillations at velocity $\mathbf{U}(t)$. The acoustic pressure is given in the far field by $p = -\rho_0 \partial \varphi / \partial t$ (see Section 1.3), and therefore

$$p(\mathbf{x}, t) = \frac{x_j}{4\pi c_0 |\mathbf{x}|^2} \frac{\partial^2 U_i}{\partial t^2} \left(t - \frac{|\mathbf{x}|}{c_0} \right) \left\{ \rho_0 \oint_S n_i y_j \, dS - \rho_0 \oint_S n_i \varphi_j^* \, dS \right\},$$

$$|\mathbf{x}| \to \infty. \quad (3.8.8)$$

The first integral is evaluated by applying the divergence theorem, which transforms it into an integral over the volume V_s of the body:

$$\rho_0 \oint_S n_i y_j \, dS = \rho_0 \int_{V_s} \frac{\partial y_j}{\partial y_i} \, d^3\mathbf{y} = \rho_0 V_s \delta_{ij} \equiv m_0 \delta_{ij}, \qquad (3.8.9)$$

where m_0 is the mass of the fluid displaced by the body. The second term in the brace brackets of (3.8.8) is just the added mass tensor M_{ij}.

Thus, the acoustic pressure can be expressed in either of the forms

$$p(\mathbf{x}, t) \approx \frac{x_i}{4\pi c_0 |\mathbf{x}|^2} (m_0 \delta_{ij} + M_{ij}) \frac{\partial^2 U_j}{\partial t^2} \left(t - \frac{|\mathbf{x}|}{c_0} \right)$$

$$= \frac{x_i}{4\pi c_0 |\mathbf{x}|^2} \left(m_0 \frac{\partial^2 U_i}{\partial t^2} + \frac{\partial F_i}{\partial t} \right) \left(t - \frac{|\mathbf{x}|}{c_0} \right), \quad |\mathbf{x}| \to \infty, \quad (3.8.10)$$

where the second line follows from (3.8.7), where F_i is the force exerted by the body on the fluid in the i direction.

For a sphere of radius a oscillating at speed $U(t)$ in the x_1 direction

$$m_0 = \tfrac{4}{3}\pi a^3 \rho_0 \quad \text{and} \quad M_{ij} = \tfrac{1}{2} m_0 \delta_{ij}$$

Therefore,

$$p(\mathbf{x}, t) \approx \frac{x_i}{4\pi c_0 |\mathbf{x}|^2} \left(m_0 \delta_{ij} + \frac{1}{2} m_0 \delta_{ij} \right) \frac{\partial^2 U_j}{\partial t^2} \left(t - \frac{|\mathbf{x}|}{c_0} \right), \quad |\mathbf{x}| \to \infty.$$

$$= \frac{\rho_0 a^3 \cos\theta}{2 c_0 |\mathbf{x}|} \frac{\partial^2 U}{\partial t^2} \left(t - \frac{|\mathbf{x}|}{c_0} \right),$$

which is equivalent to (3.8.5).

3.9 Compact Green's Function Summary and Special Cases

3.9.1 Compact Bodies and Cylindrical Bodies of Compact Cross Section

General Form

$$G(\mathbf{x}, \mathbf{y}, t - \tau) = \frac{1}{4\pi |\mathbf{X} - \mathbf{Y}|} \delta \left(t - \tau - \frac{|\mathbf{X} - \mathbf{Y}|}{c_0} \right) \qquad (3.9.1)$$

$$\left. \begin{array}{l} \mathbf{X} = \mathbf{x} - \boldsymbol{\varphi}^*(\mathbf{x}) \\[4pt] \mathbf{Y} = \mathbf{y} - \boldsymbol{\varphi}^*(\mathbf{y}) \end{array} \right\} \text{ Kirchhoff vectors for the body.}$$

The vector components $X_j(\mathbf{x})$ and $Y_j(\mathbf{y})$ are the velocity potentials of

Table 3.9.1. *Standard Special Cases*

Body	X_1	X_2	X_3
Sphere of radius a, with centre at origin	$x_1\left(1 + \frac{a^3}{2\|\mathbf{x}\|^3}\right)$	$x_2\left(1 + \frac{a^3}{2\|\mathbf{x}\|^3}\right)$	$x_3\left(1 + \frac{a^3}{2\|\mathbf{x}\|^3}\right)$
Circular cylinder of radius a coaxial with the x_3-axis	$x_1\left(1 + \frac{a^2}{x_1^2+x_2^2}\right)$	$x_2\left(1 + \frac{a^2}{x_1^2+x_2^2}\right)$	x_3
Strip airfoil $-a < x_1 < a,\ x_2 = 0,$ $-\infty < x_3 < \infty$	x_1	$\mathrm{Re}(-i\sqrt{z^2 - a^2})$ $z = x_1 + i x_2$	x_3

incompressible flow past the body having unit speed in the j direction at large distances from the body (special cases are listed in Table 3.9.1); φ_j^* is the velocity potential of the incompressible flow that would be produced by rigid body motion of S at unit speed in the j direction. For a cylindrical body of *compact cross section* parallel to the x_3 direction, we take

$$X_3 = x_3, \qquad Y_3 = y_3.$$

3.9.2 Airfoil of Variable Chord

The compact Green's function defined by (3.6.6) and (3.6.7) for a rigid strip can be generalized to include the finite span, variable chord airfoil illustrated in Fig. 3.9.1. The coordinate axes are orientated as in Fig. 3.6.3 for the strip airfoil, with y_2 normal to the plane of the airfoil and y_3 in the spanwise direction. The airfoil span is assumed to be large, and the chord $2a \equiv 2a(y_3)$ is a slowly varying function of y_3. The potential Y_2 of flow past the airfoil in the y_2 direction may then be approximated locally by the formula for an airfoil of uniform chord $2a(y_3)$.

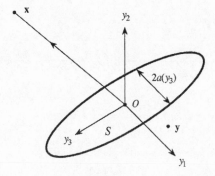

Fig. 3.9.1.

Therefore, a first approximation to the compact Green's function (3.9.1) for an airfoil of span L occupying the interval $-\frac{1}{2}L < y_3 < \frac{1}{2}L$ is obtained by taking

$$Y_1 = y_1, \qquad Y_2 = \begin{cases} \mathrm{Re}\left(-i\sqrt{z^2 - a(y_3)^2}\right), & |y_3| < \frac{1}{2}L \\ y_2, & |y_3| > \frac{1}{2}L \end{cases},$$

$$Y_3 = y_3, \qquad z = y_1 + iy_2. \tag{3.9.2}$$

This model has been found to give predictions within a few percent of those based on the exact value of $Y_2(\mathbf{y})$ in the case of an airfoil of *elliptic* planform whose

$$\text{aspect ratio} = \frac{\text{airfoil span}}{\text{midchord}} > 5.$$

3.9.3 Projection or Cavity on a Plane Wall

Let the plane wall be rigid and coincide with $x_2 = 0$ (Fig. 3.9.2). When the projection or cavity is absent the Green's function with vanishing normal derivative on the wall is

$$G_0(\mathbf{x}, \mathbf{y}, t - \tau) = \frac{1}{4\pi |\mathbf{x} - \mathbf{y}|} \delta\left(t - \tau - \frac{|\mathbf{x} - \mathbf{y}|}{c_0}\right)$$

$$+ \frac{1}{4\pi |\bar{\mathbf{x}} - \mathbf{y}|} \delta\left(t - \tau - \frac{|\bar{\mathbf{x}} - \mathbf{y}|}{c_0}\right),$$

Fig. 3.9.2.

where $\bar{\mathbf{x}} = (x_1, -x_2, x_3)$ is the image of the observer position $\mathbf{x}$ in the plane wall.

The figure illustrates the case for a projection, but the following discussion applies without change to compact (but nonresonant) wall cavities. Assume first that the origin is close to the projection. Let $|\mathbf{x}| \to \infty$ (noting that $|\bar{\mathbf{x}}| = |\mathbf{x}|$) and expand G_0 near the projection to first order in $\mathbf{y}$ (i.e., correct to dipole order)

$$G_0(\mathbf{x}, \mathbf{y}, t - \tau)$$

$$\approx \frac{1}{4\pi |\mathbf{x}|} \left\{ \delta \left(t - \tau - \frac{|\mathbf{x}|}{c_0} + \frac{\mathbf{x} \cdot \mathbf{y}}{c_0 |\mathbf{x}|} \right) + \delta \left(t - \tau - \frac{|\mathbf{x}|}{c_0} + \frac{\bar{\mathbf{x}} \cdot \mathbf{y}}{c_0 |\mathbf{x}|} \right) \right\}$$

$$\approx \frac{1}{4\pi |\mathbf{x}|} \left\{ 2\delta \left(t - \tau - \frac{|\mathbf{x}|}{c_0} \right) + \frac{2(x_1 y_1 + x_3 y_3)}{c_0 |\mathbf{x}|} \delta' \left(t - \tau - \frac{|\mathbf{x}|}{c_0} \right) \right\}.$$

We require a corrected expression that has vanishing normal derivative (as a function of $\mathbf{y}$) on the wall and on the projection. By inspection, this is obtained simply by replacing the factor

$$\frac{2(x_1 y_1 + x_3 y_3)}{c_0 |\mathbf{x}|} \quad \text{by} \quad \frac{2(x_1 Y_1 + x_3 Y_3)}{c_0 |\mathbf{x}|},$$

where $Y_1 = y_1 - \varphi_1^*(\mathbf{y})$, $Y_3 = y_3 - \varphi_3^*(\mathbf{y})$ are the velocity potentials of horizontal flows past the projection that are parallel to the wall and have unit speeds respectively in the y_1 and y_3-directions as $|\mathbf{y}| \to \infty$.

It may now be verified that (in the usual notation) the required compact Green's function is

$$G(\mathbf{x}, \mathbf{y}, t - \tau) = \frac{1}{4\pi |\mathbf{X} - \mathbf{Y}|} \delta \left(t - \tau - \frac{|\mathbf{X} - \mathbf{Y}|}{c_0} \right)$$

$$+ \frac{1}{4\pi |\bar{\mathbf{X}} - \mathbf{Y}|} \delta \left(t - \tau - \frac{|\bar{\mathbf{X}} - \mathbf{Y}|}{c_0} \right), \quad (3.9.3)$$

where

$$\left. \begin{array}{lll} Y_1 = y_1 - \varphi_1^*(\mathbf{y}), & Y_2 = y_2, & Y_3 = y_3 - \varphi_3^*(\mathbf{y}) \\ X_1 = x_1 - \varphi_1^*(\mathbf{x}), & X_2 = x_2, & X_3 = x_3 - \varphi_3^*(\mathbf{x}) \end{array} \right\}, \quad (3.9.4)$$

and $\bar{\mathbf{X}} = (X_1, -X_2, X_3)$.

These formulae can be used also for a two-dimensional projection or cavity that is uniform, say, in the x_3 direction simply by setting $Y_3 = y_3$, $X_3 = x_3$.

To complete this discussion of compact Green's function, we now give without proofs a selection of useful examples.

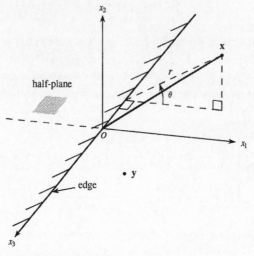

Fig. 3.9.3.

3.9.4 Green's Function for a Half-Plane (Howe, 1975a)

Analytical representations of the *exact* Green's function $\hat{G}(\mathbf{x}, \mathbf{y}, \omega)$ are known for a rigid *half-plane* $x_1 < 0, x_2 = 0$ (which is infinite in the x_3 direction, Fig. 3.9.3) but are of limited use in applications. However, we can define a compact Green's function for a source at $\mathbf{y}$ *whose distance from the edge* is small compared to the acoustic wavelength, that is, for $\kappa_0(y_1^2 + y_2^2)^{\frac{1}{2}} \ll 1$. To do this, we introduce cylindrical polar coordinates

$$\mathbf{x} = (r\cos\theta, r\sin\theta, x_3), \qquad \mathbf{y} = (r_0\cos\theta_0, r_0\sin\theta_0, y_3).$$

Then, if $\mathbf{i}_3$ is a unit vector in the x_3 direction (parallel to the edge),

$$\hat{G}(\mathbf{x}, \mathbf{y}, \omega) = \hat{G}_0(\mathbf{x}, \mathbf{y}, \omega) + \hat{G}_1(\mathbf{x}, \mathbf{y}, \omega) + \cdots, \qquad (3.9.5)$$

where, for $|\mathbf{x} - y_3\mathbf{i}_3| \to \infty$ and $\kappa_0\sqrt{y_1^2 + y_2^2} \ll 1$,

$$\hat{G}_0(\mathbf{x}, \mathbf{y}, \omega) = \frac{-1}{4\pi|\mathbf{x} - y_3\mathbf{i}_3|}e^{i\kappa_0|\mathbf{x} - y_3\mathbf{i}_3|},$$

$$\hat{G}_1(\mathbf{x}, \mathbf{y}, \omega) = \frac{-1}{\pi\sqrt{2\pi i}}\frac{\sqrt{\kappa_0}\varphi^*(\mathbf{x})\varphi^*(\mathbf{y})}{|\mathbf{x} - y_3\mathbf{i}_3|^{3/2}}e^{i\kappa_0|\mathbf{x} - y_3\mathbf{i}_3|}, \qquad (3.9.6)$$

and

$$\varphi^*(\mathbf{x}) = \sqrt{r}\sin(\theta/2), \qquad \varphi^*(\mathbf{y}) = \sqrt{r_0}\sin(\theta_0/2). \qquad (3.9.7)$$

$\varphi^*(\mathbf{x})$ is a velocity potential of incompressible flow around the edge of the half-plane expressed in terms of polar coordinates $(x_1, x_2) = r(\cos\theta, \sin\theta)$ (and similarly for $\varphi^*(\mathbf{y})$). The component $\hat{G}_0$ of $\hat{G}$ represents the radiation from a point source at $\mathbf{y}$ when scattering is neglected; $\hat{G}_1$ is the first correction due to presence of the half-plane.

3.9.5 Two-Dimensional Green's Function for a Half-Plane (Howe, 1975a)

The Green's function for the wave equation in *two dimensions*, where conditions are uniform in the x_3 direction, satisfies

$$\left(\frac{1}{c_0^2}\frac{\partial^2}{\partial t^2} - \nabla^2\right)G = \delta(x_1 - y_1)\delta(x_2 - y_2)\delta(t - \tau),$$

$$\text{where } G = 0 \quad \text{for } t < \tau. \quad (3.9.8)$$

When a line source at $\mathbf{y} = (y_1, y_2)$ is close to the edge of the half-plane in Fig. 3.9.3 the corresponding compact Green's function is obtained by integrating (3.9.6) over $-\infty < y_3 < \infty$, using the *method of stationary phase* for $\kappa_0\sqrt{x_1^2 + x_2^2} \to \infty$ (see Example 2 of Section 5.2), and then using the Formula (3.1.6) to calculate $G(\mathbf{x}, \mathbf{y}, t - \tau) \approx G_0(\mathbf{x}, t - \tau) + G_1(\mathbf{x}, \mathbf{y}, t - \tau) + \cdots$. In particular $G_1(\mathbf{x}, \mathbf{y}, t - \tau)$ is the first term in the expansion that involves $\mathbf{y}$, and is found to be

$$G_1(\mathbf{x}, \mathbf{y}, t - \tau) \approx \frac{\varphi^*(\mathbf{x})\varphi^*(\mathbf{y})}{\pi|\mathbf{x}|}\delta(t - \tau - |\mathbf{x}|/c_0), \quad |\mathbf{x}| \to \infty, \quad (3.9.9)$$

where $\mathbf{x} = (x_1, x_2)$, $\mathbf{y} = (y_1, y_2)$ and φ^* is defined as in (3.9.7).

3.9.6 Two-Dimensional Green's Function for a Plane with an Aperture

A rigid plane $x_1 = 0$ is pierced by a two-dimensional aperture occupying $-a < x_2 < a$ (Fig. 3.9.4). The two-dimensional compact Green's function (the solution of (3.9.8)) is applicable for a source at $\mathbf{y} = (y_1, y_2)$ well within an acoustic wavelength of the aperture on either side of the plane. The observer at $\mathbf{x} = (x_1, x_2)$ is in the acoustic far field. The $\mathbf{y}$-dependent part of the compact Green's function is

$$G(\mathbf{x}, \mathbf{y}, t - \tau) \approx -\frac{\sqrt{c_0}\,\mathrm{sgn}(x_1)}{\pi\sqrt{2\pi|\mathbf{x}|}}\frac{\chi(t - \tau - |\mathbf{x}|/c_0)}{\sqrt{t - \tau - |\mathbf{x}|/c_0}}\mathrm{Re}\left\{\ln\left(\frac{Z}{a} + \sqrt{\frac{Z^2}{a^2} - 1}\right)\right\},$$

$$Z = y_2 + iy_1, \quad (3.9.10)$$

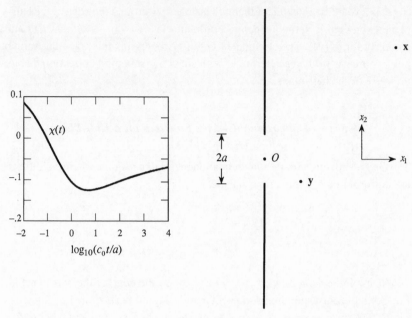

Fig. 3.9.4.

where

$$\chi(t) = H(t) \int_0^\infty \frac{\ln(\varpi a\xi^2/4c_0t)e^{-\xi^2}d\xi}{[\ln(\varpi a\xi^2/4c_0t)]^2 + \pi^2},$$

and $\varpi = 1.781072.\ldots$ Note the definition $Z = y_2 + iy_1$.

3.9.7 Green's Function for Long Waves in a Rigid Walled Duct (Howe, 1975b)

Only plane waves can propagate in a cylindrical duct of cross-sectional area A when the characteristic wavelength of sound is large compared with the diameter $\sim\sqrt{A}$, even if the source region is highly three dimensional. When this condition is satisfied the corresponding compact Green's function satisfies the one-dimensional wave equation, provided the cross-sectional area is *uniform*. Taking the x_1 direction along the axis of the duct (Fig. 3.9.5a), we have

$$G(\mathbf{x}, \mathbf{y}, t - \tau) = \frac{c_0}{2A} H\left(t - \tau - \frac{|x_1 - y_1|}{c_0}\right), \quad |\mathbf{x} - \mathbf{y}| \gg \sqrt{A}, \quad (3.9.11)$$

where H is the Heaviside step function. For a uniform duct with an acoustically

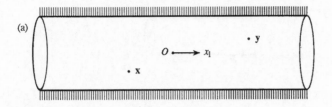

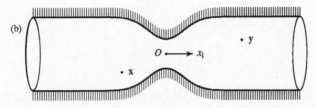

Fig. 3.9.5.

compact section of variable cross section, such as the neck in Fig. 3.9.5b, the compact Green's function becomes

$$G(\mathbf{x}, \mathbf{y}, t - \tau) = \frac{c_0}{2A} H\left(t - \tau - \frac{|X_1 - Y_1|}{c_0}\right), \quad |\mathbf{x} - \mathbf{y}| \gg \sqrt{A}, \quad (3.9.12)$$

where $X_1(\mathbf{x})$ and $Y_1(\mathbf{y})$ are the velocity potential of incompressible flow in the duct having unit speed at large distances from the neck.

3.9.8 Compact Green's Function for a Duct Entrance (Howe, 1998a,b)

The typical geometry is illustrated in Fig. 3.9.6a. Within the duct, several diameters from the entrance, the cross-sectional area is uniform and equal to A. However, the geometry of the entrance can be arbitrary, and not necessarily that of the uniform cylinder shown in the figure. Take the coordinate origin in the entrance plane of the duct, with the negative x_1 axis lying along the axis of the duct. Then, there are two cases:

(i) Propagation within the Duct

This is applicable for the case shown in Fig. 3.9.6a, involving a source at $\mathbf{y}$ near the duct entrance and an observer at $\mathbf{x}$ within the duct (or vice versa), when the

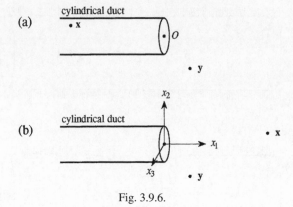

Fig. 3.9.6.

characteristic acoustic wavelength is large compared to the duct diameter.

$$G(\mathbf{x}, \mathbf{y}, t - \tau) \approx \frac{c_0}{2A} \left\{ H\left[t - \tau - \frac{|\varphi^*(\mathbf{x}) - \varphi^*(\mathbf{y})|}{c_0} \right] \right.$$
$$\left. - H\left[t - \tau + \frac{\varphi^*(\mathbf{x}) + \varphi^*(\mathbf{y})}{c_0} \right] \right\}, \qquad (3.9.13)$$

where the velocity potential $\varphi^*(\mathbf{y})$ describes incompressible flow from the duct, and satisfies

$$\varphi^*(\mathbf{x}) \approx x_1 - \ell' \qquad \text{when } |x_1| \gg \sqrt{A} \text{ within the duct,}$$
$$\approx -A/4\pi|\mathbf{x}| \quad \text{when } |\mathbf{x}| \gg \sqrt{A} \text{ outside the duct,} \qquad (3.9.14)$$

in which ℓ' is the end correction (Rayleigh, 1945) of the duct opening ($\approx 0.61R$ for an unflanged circular cylinder of radius $R = \sqrt{A/\pi}$).

(ii) Propagation in Free Space (Fig. 3.9.6b)

When either the source or observer is located at a large distance from the duct entrance in *free space*

$$G(\mathbf{x}, \mathbf{y}, t - \tau) = \frac{1}{4\pi|\mathbf{X} - \mathbf{Y}|} \delta\left(t - \tau - \frac{(|\mathbf{X}(\mathbf{x}) - \mathbf{Y}(\mathbf{y})| - [\varphi^*(\mathbf{x}) + \varphi^*(\mathbf{y})])}{c_0} \right),$$
$$(3.9.15)$$

where φ^* is defined as in (3.9.14), and $\mathbf{X}(\mathbf{x})$, $\mathbf{Y}(\mathbf{y})$ denote the Kirchhoff vector whose i component is the velocity potential of flow past the stationary surface formed by the duct entrance having unit speed in the i direction at large distances from the entrance outside the duct (they become exponentially small with distance $|x_1|$ or $|y_1|$ *into* the duct). The terms $\varphi^*(\mathbf{x})$, $\varphi^*(\mathbf{y})$ account for the additional, weak *monopole* sound generated by a source near the duct entrance; the source compresses the fluid in the duct mouth producing a sound wave in the duct whose reaction on the mouth causes a volume flux equal to the monopole source strength. The amplitude of this monopole is of the same order as the usual dipole sound determined by the compact Green's function.

For a uniform, thin-walled cylindrical duct we can take

$$\mathbf{X}(\mathbf{x}) \equiv (x_1 - \varphi^*(\mathbf{x}), X_2(\mathbf{x}), X_3(\mathbf{x})), \qquad \mathbf{Y}(\mathbf{y}) \equiv (y_1 - \varphi^*(\mathbf{y}), Y_2(\mathbf{y}), Y_3(\mathbf{y})).$$

If the source coordinate $y_1 \to -\infty$ within the duct,

$$G(\mathbf{x}, \mathbf{y}, t - \tau) = \frac{1}{4\pi |\mathbf{x}|} \delta(t - \tau - (|\mathbf{x}| - y_1)/c_0), \quad |\mathbf{x}| \to \infty. \qquad (3.9.16)$$

This represents a monopole wave centered on the duct entrance. This limiting form of G can be used to calculate the low-frequency free space radiation generated by internal sources far from the entrance.

Problems 3

1. Use the compact Green's function to solve

$$\left(\nabla^2 + \kappa_0^2\right) \hat{\varphi} = f_2 \frac{\partial}{\partial x_2} \left[\delta(x_1 - L)\delta(x_2)\delta(x_3)\right],$$

where

$$\frac{\partial \varphi}{\partial x_\mathrm{n}} = 0 \quad \text{on} \quad |\mathbf{x}| = a.$$

for the sound radiated by an azimuthally orientated dipole adjacent to a compact, rigid sphere.

2. Use the compact Green's function to solve

$$\left(\nabla^2 + \kappa_0^2\right) \hat{\varphi} = f_1 \frac{\partial}{\partial x_1} [\delta(x_1 - L)\delta(x_2)\delta(x_3)],$$

where

$$\frac{\partial \varphi}{\partial x_\mathrm{n}} = 0 \quad \text{on} \quad \left(x_1^2 + x_2^2\right)^{\frac{1}{2}} = a.$$

for the sound radiated by a radially orientated dipole adjacent to a rigid circular cylinder of compact cross section.

3. Repeat Question 2 for the dipoles

$$f_2 \frac{\partial}{\partial x_2}[\delta(x_1 - L)\delta(x_2)\delta(x_3)], \qquad f_3 \frac{\partial}{\partial x_3}[\delta(x_1 - L)\delta(x_2)\delta(x_3)].$$

4. The constant strength dipole

$$q(\mathbf{x}, t) = f_2 \frac{\partial}{\partial x_2}(\delta(x_1 - Ut)\delta(x_2 - h)\delta(x_3)), \qquad f_2 = \text{constant}$$

translates at *constant* velocity U past a fixed rigid cylinder of radius $a < h$ whose axis coincides with the x_3 axis. Show that when $M = U/c_0 \ll 1$, the far-field acoustic potential determined by Equation (3.1.1) is given by

$$\varphi \approx \frac{f_2 M a^2}{2\pi |\mathbf{x}|(h^2 + U^2[t]^2)^3} \left\{ \frac{x_1 h}{|\mathbf{x}|}(3U^2[t]^2 - h^2) + \frac{x_2 U[t]}{|\mathbf{x}|}(3h^2 - U^2[t]^2) \right\},$$

where

$$[t] = t - \frac{|\mathbf{x}|}{c_0}.$$

5. The volume source

$$q(\mathbf{x}, t) = q_0 \delta(x_1 - Ut)\delta(x_2 - h)\delta(x_3), \qquad q_0 = \text{constant}$$

translates at *constant* velocity U past a fixed rigid sphere of radius $a < h$ whose center is at the origin. Determine from Equation (3.1.1) the far field acoustic pressure $p = -\rho_0 \partial\varphi/\partial t$ given that $M = U/c_0 \ll 1$.

6. The point source $q(\mathbf{x}, t) = q_0 \delta(x_1 - Ut)\delta(x_2)\delta(x_3)$, ($q_0 = $ constant) convects along the axis of symmetry of the necked duct shown in Fig. 3.9.5 at constant, low Mach number speed U. If the cross-sectional area of the duct is denoted by $S(x_1)$, where $S(x_1) \rightarrow A$, $x_1 \rightarrow \pm\infty$, use the approximations

$$X_1 = A \int_0^{x_1} \frac{d\xi}{S(\xi)}, \qquad Y_1 = A \int_0^{y_1} \frac{d\xi}{S(\xi)},$$

to calculate the acoustic pressure radiated from the neck during the passage of the source.

7. In incompressible flow the velocity potential generated by a distribution of sources $q(\mathbf{x}, t)$ near a rigid body is determined by the solution of

$$\nabla^2 \varphi = q(\mathbf{x}, t).$$

Show that the monopole and dipole components of the solution at large distances from the body (in the *hydrodynamic* far field) can be calculated using the following incompressible limit of the compact Green's function

$$G(\mathbf{x}, \mathbf{y}, t - \tau) = \frac{-\delta(t - \tau)}{4\pi |\mathbf{X} - \mathbf{Y}|}.$$

8. A rigid body translates without rotation in the j direction at velocity $U_j(t)$ in an ideal, incompressible fluid at rest at infinity. Show that the velocity potential of the fluid motion is

$$\Phi = U_j \varphi_j^*.$$

In a fixed reference frame, the pressure can be calculated from Bernoulli's equation:

$$\frac{\partial \Phi}{\partial t} + \frac{p}{\rho_0} + \frac{1}{2}(\nabla \Phi)^2 = 0.$$

Use these results to prove formula (3.8.7) for the force exerted on the fluid by the body.

9. A compact rigid disc of radius a executes small amplitude vibrations at velocity $U(t)$ normal to itself. In the undisturbed state it lies in the plane $x_1 = 0$ with its center at the origin. If $\varphi_1^*(\mathbf{x}) = \mp(2/\pi)\sqrt{a^2 - x_2^2 - x_3^2}$ on the faces $x_1 = \pm 0$, $\sqrt{x_2^2 + x_3^2} < a$ of the disc, show that the acoustic pressure generated by the motion is given by

$$p(\mathbf{x}, t) \approx \frac{2\rho_0 a^3 \cos\theta}{3\pi c_0 |\mathbf{x}|} \frac{\partial^2 U}{\partial t^2}(t - |\mathbf{x}|/c_0), \quad |\mathbf{x}| \to \infty, \quad \cos\theta = \frac{x_1}{|\mathbf{x}|}.$$

4

Vorticity

4.1 Vorticity and the Kinetic Energy of Incompressible Flow

4.1.1 Kelvin's (1867) Definition

Kelvin was responsible for much of the pioneering work on the mechanics of incompressible flow. He gave the following definition of a **vortex** in a homogeneous incompressible fluid,

...a portion of fluid having any motion that it could not acquire by fluid pressure transmitted from its boundary.

To understand this consider the ideal (i.e., inviscid) incompressible flow produced by arbitrary motion of a solid body with surface S (Fig. 4.1.1). The motion generated from rest by 'fluid pressure transmitted from its boundary' can be described by a velocity potential φ such that

$$\mathbf{v}(\mathbf{x}, t) = \nabla\varphi, \qquad \frac{\partial\varphi}{\partial x_n} = U_n \quad \text{on } S,$$

where U_n is the normal component of velocity on S.

There are no sources within the instantaneous region V occupied by the fluid, where $\nabla^2\varphi = 0$. Therefore, the kinetic energy T_0 of the flow is

$$T_0 = \frac{1}{2}\rho_0 \int_V (\nabla\varphi)^2 \, d^3\mathbf{x} = \frac{1}{2}\rho_0 \int_V (\text{div}(\varphi\nabla\varphi) - \varphi\nabla^2\varphi) \, d^3\mathbf{x}$$

$$= -\frac{1}{2}\rho_0 \oint_S \varphi\frac{\partial\varphi}{\partial x_n} \, dS \equiv -\frac{1}{2}\rho_0 \oint_S \varphi U_n \, dS, \tag{4.1.1}$$

where the divergence theorem has been used to obtain the second line (there is no contribution from the surface Σ at infinity in Fig. 4.1.1, where $\varphi \sim O(1/|\mathbf{x}|^2)$). This formula implies that if S is suddenly brought to rest ($U_n \to 0$) the motion everywhere in the fluid ceases instantaneously, because $\int_V (\nabla\varphi)^2 \, d^3\mathbf{x}$

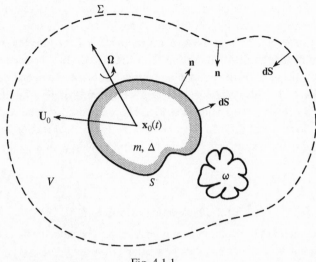

Fig. 4.1.1.

can vanish only if $\nabla \varphi \equiv 0$. This unphysical behavior is never observed in a real fluid because (i) no fluid is perfectly incompressible, and signals generated by changes in the boundary conditions propagate at the finite speed of sound, and (ii) diffusion of *vorticity* from the boundary supplies irrecoverable kinetic energy to the fluid.

For an incompressible, real fluid we write

$$\mathbf{v} = \nabla \varphi + \mathbf{u},$$

and define the *vorticity* $\boldsymbol{\omega}$ by

$$\boldsymbol{\omega} = \operatorname{curl} \mathbf{u} \equiv \operatorname{curl} \mathbf{v}.$$

If φ is taken to be defined as above (for ideal flow) then, because div $\mathbf{u} = 0$ and the normal component $u_n = \mathbf{n} \cdot \mathbf{u} = 0$ on S, the kinetic energy becomes

$$
\begin{aligned}
T &= \frac{1}{2}\rho_0 \int_V (\nabla \varphi + \mathbf{u})^2 \, d^3\mathbf{x} = \frac{1}{2}\rho_0 \int_V ((\nabla \varphi)^2 + 2\nabla \varphi \cdot \mathbf{u} + u^2) \, d^3\mathbf{x} \\
&= \frac{1}{2}\rho_0 \int_V ((\nabla \varphi)^2 + u^2) \, d^3\mathbf{x} + \rho_0 \int_V \operatorname{div}(\varphi \mathbf{u}) \, d^3\mathbf{x} \\
&= \frac{1}{2}\rho_0 \int_V ((\nabla \varphi)^2 + u^2) \, d^3\mathbf{x} - \rho_0 \oint_S \varphi u_n \, dS \\
&= -\frac{1}{2}\rho_0 \oint_S \varphi U_n \, dS + \frac{1}{2}\rho_0 \int_V u^2 \, d^3\mathbf{x} \equiv T_0 + \frac{1}{2}\rho_0 \int_V u^2 \, d^3\mathbf{x}. \quad (4.1.2)
\end{aligned}
$$

When the surface motion is arrested, the flow described by the velocity potential φ stops instantaneously, but that associated with the rotational velocity **u** persists. The crucial difference between rotational and irrotational flows is that, once established, vortical motions proceed irrespective of whether or not the fluid continues to be driven by moving boundaries or other external agencies.

Equation (4.1.2) also establishes **Kelvin's theorem** that $T \geq T_0$: the kinetic energy of the real flow (for which $\mathbf{u} \neq 0$) always exceeds that of the corresponding ideal, irrotational flow. In other words, the irrotational motion represents the *least* possible disturbance that can be produced in the fluid by the moving boundary.

4.2 The Vorticity Equation

Let $\mathbf{v}_A$ denote the fluid velocity at a point A at **x**. The velocity $\mathbf{v}_B$ at a neighbouring point B at $\mathbf{x} + \delta\mathbf{x}$ can then be written (Goldstein, 1960)

$$\mathbf{v}_B \approx \mathbf{v}_A + (\delta\mathbf{x} \cdot \nabla)\mathbf{v}$$
$$= \mathbf{v}_A + \tfrac{1}{2}\boldsymbol{\omega} \wedge \delta\mathbf{x} + \tfrac{1}{2}\nabla(e_{ij}\delta x_i \delta x_j),$$

where e_{ij} is the rate-of-strain tensor (2.1.3) and the gradient in the second line is taken with respect to $\delta\mathbf{x}$. The first two terms on the second line represent motion of A and B *as a rigid body,* consisting of a translation at velocity $\mathbf{v}_A$ together with a rotation at angular velocity $\tfrac{1}{2}\boldsymbol{\omega}$; the last term, being a gradient, represents an irrotational distortion of the fluid in the neighborhood of A.

If we consider a small spherical fluid particle with center at A, the distortion corresponds to a deformation into an ellipsoid whose principal axes correspond to the principal axes of e_{ij}. If a spherical fluid particle is suddenly solidified without change of angular momentum, it will rotate at angular velocity $\boldsymbol{\omega}/2$, so that $\boldsymbol{\omega}$ may be defined as twice the initial angular velocity of the solid sphere when an infinitesimally small sphere of fluid with center at A is suddenly solidified without change of angular momentum (but this is not true for arbitrarily shaped volume elements). The vorticity may therefore be regarded as a measure of the angular momentum of a fluid particle. This is consistent with our conclusion above regarding kinetic energy, inasmuch as the conservation of angular momentum suggests that vorticity is associated the *intrinsic* kinetic energy of the flow, and determines the motion that persists in an incompressible fluid when the boundaries are brought to rest.

A **vortex line** is tangential to the vorticity vector at all points along its length. Vortex lines that pass through every point of a simple closed curve define the boundary of a *vortex tube*. For a tube of small cross-sectional area δS the product

$\omega\,\delta S$ is called the tube strength, and is constant because

$$\operatorname{div}\omega = \operatorname{div}(\operatorname{curl}\mathbf{v}) \equiv 0,$$

and the divergence theorem therefore implies that $\oint \omega \cdot d\mathbf{S} = 0$ for any closed surface, and in particular for the surface formed by two cross sections of the tube and the tube wall separating them, on the latter of which $\omega \cdot d\mathbf{S} = 0$. It follows that vortex tubes and lines cannot begin or end within the fluid. The no-slip condition (Batchelor 1967) requires the velocity at a boundary to be the same as that of the boundary. A vortex line must therefore form a closed loop, or end on a *rotating* surface S at which

$$\mathbf{n}\cdot\omega = 2\mathbf{n}\cdot\Omega, \tag{4.2.1}$$

where Ω is the angular velocity of S.

We shall show that vorticity is transported by convection and molecular diffusion. Therefore an initially confined region of vortex loops can frequently be assumed to remain within a bounded region. In the absence of body forces $\mathbf{F}$, we first use the identity $\operatorname{curl}\operatorname{curl}\mathbf{A} = \operatorname{grad}\operatorname{div}\mathbf{A} - \nabla^2\mathbf{A}$ to write the momentum equation (1.2.3) for homentropic flow in the form

$$\frac{\partial\mathbf{v}}{\partial t} + (\mathbf{v}\cdot\nabla)\mathbf{v} + \nabla\left(\int\frac{dp}{\rho}\right) = -\nu\left(\operatorname{curl}\omega - \frac{4}{3}\nabla(\operatorname{div}\mathbf{v})\right).$$

By using the vector identity

$$(\mathbf{v}\cdot\nabla)\mathbf{v} = \omega\wedge\mathbf{v} + \nabla\left(\tfrac{1}{2}v^2\right), \tag{4.2.2}$$

the momentum equation can be cast into **Crocco's form**

$$\frac{\partial\mathbf{v}}{\partial t} + \omega\wedge\mathbf{v} + \nabla B = -\nu\left(\operatorname{curl}\omega - \frac{4}{3}\nabla(\operatorname{div}\mathbf{v})\right), \tag{4.2.3}$$

where

$$B = \int\frac{dp}{\rho} + \frac{1}{2}v^2 \tag{4.2.4}$$

is the **total enthalpy** in homentropic flow. The vector $\omega\wedge\mathbf{v}$ is sometimes called the **Lamb vector**. When the fluid is incompressible (or when the term in div $\mathbf{v}$ representing the small effect of compressibility on *viscous dissipation* is neglected) Crocco's equation reduces to

$$\frac{\partial\mathbf{v}}{\partial t} + \omega\wedge\mathbf{v} + \nabla B = -\nu\operatorname{curl}\omega, \tag{4.2.5}$$

in which case dissipation occurs only where $\omega \neq \mathbf{0}$.

The curl of (4.2.3) and the relation curl curl $\omega \equiv -\nabla^2\omega$ yield the **vorticity equation** for a Stokesian fluid of constant shear viscosity:

$$\frac{\partial\omega}{\partial t} + \text{curl}(\omega \wedge \mathbf{v}) = \nu\nabla^2\omega. \tag{4.2.6}$$

For an incompressible fluid div $\omega =$ div $\mathbf{v} = 0$, and

$$\text{curl}(\omega \wedge \mathbf{v}) \equiv (\mathbf{v} \cdot \nabla)\omega + \omega\,\text{div}\,\mathbf{v} - (\omega \cdot \nabla)\mathbf{v} - \mathbf{v}\,\text{div}\,\omega = (\mathbf{v} \cdot \nabla)\omega - (\omega \cdot \nabla)\mathbf{v},$$

therefore the vorticity equation can also be written

$$\frac{D\omega}{Dt} = (\omega \cdot \nabla)\mathbf{v} + \nu\nabla^2\omega. \tag{4.2.7}$$

The terms on the right represent the mechanisms that change the vorticity of a moving fluid particle in incompressible flow:

(i) $(\omega \cdot \nabla)\mathbf{v}$.

Consider a fluid particle at A in Fig. 4.2.1 with velocity $\mathbf{v}$ at time t. Let the vorticity at A be $\omega = \omega\mathbf{n}$, where $\mathbf{n}$ is a unit vector, and consider a neighboring particle at B a small distance s from A in the direction $\mathbf{n}$; that is, $s\mathbf{n}$ is the position of B relative to A. At time t the points A and B lie on the vortex line through A, and the velocity at B is $\mathbf{v} + s(\mathbf{n} \cdot \nabla)\mathbf{v}$. After a short time δt, A has moved a vector distance $\mathbf{v}\,\delta t$ to A' and B has moved to B' whose position relative to A' is $s(\mathbf{n}+(\mathbf{n} \cdot \nabla)\mathbf{v}\,\delta t)$. During this time, the term $(\omega \cdot \nabla)\mathbf{v}$ in the vorticity equation causes the vorticity of the fluid particle initially at A to change from $\omega\mathbf{n}$ at A to $\omega(\mathbf{n} + (\mathbf{n} \cdot \nabla)\mathbf{v}\,\delta t)$ at A'. Thus, the vortex line through A' lies along the relative vector $s(\mathbf{n} + (\mathbf{n} \cdot \nabla)\mathbf{v}\,\delta t)$ from A' to B'. Therefore, the fluid particles and the vortex line through A and B have deformed and convected in the flow in the

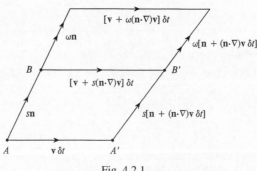

Fig. 4.2.1.

same way; in their new positions A and B continue to lie on the same vortex line. In the absence of viscosity (when $\nu\nabla^2\omega$ does not appear on the right of (4.2.7)) vortex lines are therefore said to move with the fluid; they are rotated and stretched in a manner determined entirely by the relative motions of A and B. The magnitude of ω increases in direct proportion to the stretching of vortex lines. When a vortex tube is stretched, the cross-sectional area δS decreases and therefore ω must increase to preserve the strength of the tube.

(ii) $\nu\nabla^2\omega$: Molecular Diffusion of Vorticity

This term is important only in regions of high shear, in particular near solid boundaries. Very close to a stationary wall the velocity becomes small and non-linear terms in the vorticity equation (4.2.7) can be neglected. The equation then reduces to the classical diffusion equation

$$\frac{\partial\omega}{\partial t} = \nu\nabla^2\omega.$$

Vorticity is generated at solid boundaries, and viscosity is responsible for its diffusion into the body of the fluid, where it can subsequently be convected by the flow.

It should be understood that viscosity merely serves to diffuse the vorticity into the fluid from the surface, and does not generate the vorticity. In an ideal fluid the slipping of the flow over the surface creates a singular layer of vorticity at the surface called a **vortex sheet** whose strength is determined by the tangential velocity difference between the surface and the ideal exterior flow. This vorticity stays on the surface; it would start to diffuse into the fluid if the fluid were suddenly endowed with viscosity. The *rate* of diffusion would depend on the value of ν, but the *amount* of the vorticity available for diffusion from the surface is independent of ν.

The **circulation** Γ with respect to a closed material contour C is defined by

$$\Gamma = \oint_C \mathbf{v}\cdot d\mathbf{x} = \int_{S'} \operatorname{curl}\mathbf{v}\cdot d\mathbf{S} \equiv \int_{S'} \omega\cdot d\mathbf{S},$$

where S' is any two-sided surface bounded by C. When $\nu = 0$ the motion in homentropic flow evolves in such a way that the circulation around the moving contour remains constant:

$$\frac{D\Gamma}{Dt} = \frac{D}{Dt}\oint_C \mathbf{v}\cdot d\mathbf{x} = \oint_C \nabla\left(-\int\frac{dp}{\rho} + \frac{1}{2}v^2\right)\cdot d\mathbf{x} \equiv 0. \quad (4.2.8)$$

This is **Kelvin's circulation theorem.** It follows that vorticity can neither be created nor destroyed in a body of inviscid and homentropic fluid.

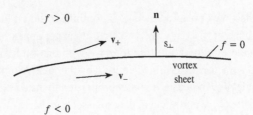

Fig. 4.2.2.

4.2.1 Vortex Sheets

A vortex sheet is a useful model of a thin layer of vorticity when viscous diffusion can be neglected. Imagine a thin shear layer (Fig. 4.2.2) across which the velocity changes rapidly from $\mathbf{v}_-$ to $\mathbf{v}_+$. We approximate the layer by a surface $f(\mathbf{x}, t) = 0$ with unit normal $\mathbf{n}$ across which the normal components of velocity are equal ($\mathbf{n} \cdot \mathbf{v}_- = \mathbf{n} \cdot \mathbf{v}_+$), but the tangential components are discontinuous ($\mathbf{n} \wedge \mathbf{v}_- \neq \mathbf{n} \wedge \mathbf{v}_+$). Let $f \gtrless 0$ respectively on the $\pm$ sides of the surface. Near the sheet, on either side, it can be assumed that curl $\mathbf{v}_\pm = \mathbf{0}$, and we can set

$$\mathbf{v} = H(f)\mathbf{v}_+ + H(-f)\mathbf{v}_-$$

Hence,

$$\boldsymbol{\omega} = \nabla H \wedge (\mathbf{v}_+ - \mathbf{v}_-) = \mathbf{n} \wedge (\mathbf{v}_+ - \mathbf{v}_-)\delta(s_\perp), \qquad (4.2.9)$$

where $\nabla H \equiv \nabla H(f) = -\nabla H(-f) = \mathbf{n}\delta(s_\perp)$, and $s_\perp$ is distance measured in the normal direction from the sheet.

In a real fluid the vorticity would diffuse out from the sheet and it could not therefore persist indefinitely. In an ideal fluid the sheet is subject only to convection and stretching by the flow at the local mean velocity, which is

$$\mathbf{v} = \tfrac{1}{2}(\mathbf{v}_+ + \mathbf{v}_-), \qquad (4.2.10)$$

where $\mathbf{v}_\pm$ are evaluated just above and below the sheet.

4.3 The Biot–Savart Law

In an unbounded fluid the velocity $\mathbf{v}$ can always be expressed in terms of scalar and vector potentials φ and $\mathbf{A}$ such that

$$\mathbf{v} = \nabla\varphi + \text{curl}\,\mathbf{A}, \quad \text{where div}\,\mathbf{A} = 0.$$

The equations determining φ and $\mathbf{A}$ are found by taking in turn the divergence and curl (using the formula curl curl $\mathbf{A} = \operatorname{grad} \operatorname{div} \mathbf{A} - \nabla^2 \mathbf{A}$):

$$\nabla^2 \varphi = \operatorname{div} \mathbf{v}, \qquad \nabla^2 \mathbf{A} = -\operatorname{curl} \mathbf{v} \equiv -\boldsymbol{\omega}.$$

We can take $\varphi = 0$ for incompressible, unbounded flow which is at rest at infinity. To find $\mathbf{A}$ we use the Green's function for Laplace's equation determined by (1.4.6) (i.e., by the incompressible limit of (3.2.6), when $\kappa_0 = 0$) to obtain

$$\mathbf{A} = \int \frac{\boldsymbol{\omega}(\mathbf{y}, t) \, d^3\mathbf{y}}{4\pi |\mathbf{x} - \mathbf{y}|}.$$

The velocity is then given by the **Biot–Savart** formula

$$\mathbf{v}(\mathbf{x}, t) = \operatorname{curl} \int \frac{\boldsymbol{\omega}(\mathbf{y}, t) \, d^3\mathbf{y}}{4\pi |\mathbf{x} - \mathbf{y}|}. \tag{4.3.1}$$

This is a purely *kinematic* relation between a vector $\mathbf{v}$ that vanishes at infinity and $\boldsymbol{\omega} = \operatorname{curl} \mathbf{v}$.

Because vorticity is transported by convection and diffusion, an initially confined region of vorticity will tend to remain within a bounded domain, so that it may be assumed that $\boldsymbol{\omega} \to 0$ as $|\mathbf{x}| \to \infty$. The divergence theorem then shows that

$$\int \omega_i(\mathbf{y}, t) \, d^3\mathbf{y} = -\oint_\Sigma y_i \omega_j(\mathbf{y}, t) n_j \, dS(\mathbf{y}) \equiv 0,$$

where the surface Σ (with inward normal $\mathbf{n}$) is large enough to contain all the vorticity. By using this result and the expansion (1.9.3) for $|\mathbf{x}| \to \infty$ we derive from (4.3.1) the following approximation in the *hydrodynamic far field:*

$$\mathbf{v}(\mathbf{x}, t) \approx \operatorname{curl} \left(\frac{x_j}{4\pi |\mathbf{x}|^3} \int y_j \boldsymbol{\omega}(\mathbf{y}, t) \, d^3\mathbf{y} \right) \sim \mathrm{O}\left(\frac{1}{|\mathbf{x}|^3} \right), \quad |\mathbf{x}| \to \infty. \tag{4.3.2}$$

Furthermore, the divergence theorem also implies that

$$\int \operatorname{div}(y_i y_j \boldsymbol{\omega}(\mathbf{y}, t)) \, d^3\mathbf{y} = 0,$$

and therefore that $\quad \int (y_i \omega_j(\mathbf{y}, t) + y_j \omega_i(\mathbf{y}, t)) \, d^3\mathbf{y} = 0.$

This can be used to express (4.3.2) in either of the following equivalent forms

$$\mathbf{v}(\mathbf{x}, t) \approx \operatorname{curl} \operatorname{curl}\left(\frac{\mathbf{I}}{4\pi |\mathbf{x}|} \right) = \operatorname{grad} \operatorname{div}\left(\frac{\mathbf{I}}{4\pi |\mathbf{x}|} \right), \quad |\mathbf{x}| \to \infty,$$

$$\text{where} \quad \mathbf{I} = \frac{1}{2} \int \mathbf{y} \wedge \boldsymbol{\omega}(\mathbf{y}, t) \, d^3\mathbf{y}, \tag{4.3.3}$$

(see Question 2 of Problems 4). The vector $\mathbf{I}$ is called the *impulse* of the vortex system, and is an *absolute constant* in an unbounded flow (see Section 4.4). These formulae supply alternative representations of $\mathbf{v}$ in the *hydrodynamic* far field (where the motion is entirely irrotational) in terms of either the vector potential $\mathbf{A} = \operatorname{curl}(\mathbf{I}/4\pi\,|\mathbf{x}|)$ or the scalar potential $\varphi = \operatorname{div}(\mathbf{I}/4\pi\,|\mathbf{x}|)$. (Batchelor (1967) denotes $\mathbf{I}$ by $\mathbf{P}/\rho_0$; Lighthill (1978, 1986) uses $\mathbf{G}$.)

4.3.1 Kinetic Energy

Using the Biot–Savart formula it can be verified that the kinetic energy of an unbounded (three-dimensional) incompressible flow is given in terms of the vorticity by

$$T = \frac{\rho_0}{8\pi} \iint \frac{\boldsymbol{\omega}(\mathbf{x}, t) \cdot \boldsymbol{\omega}(\mathbf{y}, t)}{|\mathbf{x} - \mathbf{y}|}\, d^3\mathbf{x}\, d^3\mathbf{y}. \tag{4.3.4}$$

The following representation can also be derived (using (4.2.2))

$$T = \rho_0 \int \mathbf{x} \cdot (\boldsymbol{\omega} \wedge \mathbf{v})(\mathbf{x}, t)\, d^3\mathbf{x}. \tag{4.3.5}$$

4.3.2 Incompressible Flow with an Internal Boundary

Let the rigid body in Fig. 4.1.1 have volume Δ and move in an incompressible fluid at rest at infinity with velocity

$$\mathbf{U} = \mathbf{U}_0 + \boldsymbol{\Omega} \wedge (\mathbf{x} - \mathbf{x}_0(t)), \tag{4.3.6}$$

where $\mathbf{U}_0 = d\mathbf{x}_0/dt$ is the velocity of its center of volume $\mathbf{x}_0(t)$, and $\boldsymbol{\Omega}(t)$ is its angular velocity.

In the usual way let $f(\mathbf{x}, t)$ vanish on S, with $f > 0$ in the fluid. Then $H(f)\mathbf{v} + H(-f)\mathbf{U}$ is the velocity everywhere, in both the fluid and solid (where it equals $\mathbf{U}(\mathbf{x}, t)$). But

$$H(f)\mathbf{v} + H(-f)\mathbf{U} = \nabla\varphi + \operatorname{curl}\mathbf{A}, \quad (\operatorname{div}\mathbf{A} = 0).$$

The body has constant volume ($\operatorname{div}\mathbf{U} = 0$), but $\operatorname{curl}\mathbf{U} = 2\boldsymbol{\Omega}$. Now, the no-slip condition on S implies that

$$\operatorname{div}(H(f)\mathbf{v} + H(-f)\mathbf{U}) = \nabla H(f) \cdot (\mathbf{v} - \mathbf{U}) \equiv 0$$

$$\operatorname{curl}(H(f)\mathbf{v} + H(-f)\mathbf{U}) = H(f)\boldsymbol{\omega} + H(-f)2\boldsymbol{\Omega}.$$

Hence, $\varphi \equiv 0$, and the velocity *everywhere* is given by the following modification of the Biot–Savart formula (4.3.1):

$$\mathbf{v}(\mathbf{x}, t) = \text{curl} \int_V \frac{\boldsymbol{\omega}(\mathbf{y}, t)\, d^3\mathbf{y}}{4\pi\,|\mathbf{x} - \mathbf{y}|} + \text{curl} \int_\Delta \frac{2\boldsymbol{\Omega}(t)\, d^3\mathbf{y}}{4\pi\,|\mathbf{x} - \mathbf{y}|}, \qquad (4.3.7)$$

where V is the volume occupied by the fluid. This formula predicts that $\mathbf{v} = \mathbf{U}$ when $\mathbf{x}$ lies in the region Δ occupied by the body. Vortex lines may be imagined to continue into the solid. As for an unbounded flow, the identity $\int \text{curl}\,(H(f)\mathbf{v} + H(-f)\mathbf{U})\, d^3\mathbf{x} = 0$ implies that $\mathbf{v} \sim O(1/|\mathbf{x}|^3)$ as $|\mathbf{x}| \to \infty$. Similarly, the asymptotic representations (4.3.3) remain valid provided the integration includes the region occupied by the body (where $\boldsymbol{\omega} = 2\boldsymbol{\Omega}$). The formula is also applicable in inviscid flow, but the contribution from the bound vorticity in the vortex sheet on the surface of the body must be included in the integrals.

4.3.3 Blowing Out a Candle (Lighthill 1963)

An amusing illustration of the significance of vorticity is depicted in Fig. 4.3.1, where a puff of air is ejected from the tube and directed at the flame of a candle. Suppose the tube has radius R, and that the air is forced out at constant speed V by impulsive movement of the piston over an axial distance L. In an ideal fluid the motion outside is irrotational and resembles at large distances from the exit a radially symmetric source flow. This flow persists only while the piston is moving, during which time the velocity potential at a large distance r from the

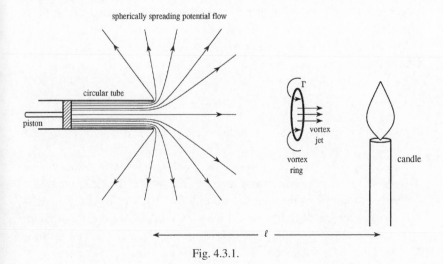

Fig. 4.3.1.

exit resembles that produced by a monopole of strength $q = \pi R^2 V$:

$$\varphi \sim -\frac{V R^2}{4r},$$

so that the air blows against the flame at distance ℓ at speed

$$V_\varphi \sim \frac{V}{4}\frac{R^2}{\ell^2}.$$

In reality, vorticity is generated at the tube wall. The air leaves the tube in the form of a jet, and the exiting fluid is initially contained within a cylindrical slug of air of length L, whose displacement from the tube forces the potential flow φ. In addition, however, vorticity leaves the tube within a circular cylindrical *vortex sheet* at the periphery of the slug, across which the axial velocity changes from V within the jet to 0 outside. The sheet may be pictured as a succession of vortex rings of radius R and infinitesimal core radii. The circulation of these rings per unit length of the jet is V, and they translate at the local mean air velocity on the sheet equal to $\frac{1}{2}V$. The total circulation ejected from the tube during the time L/V in which the piston moves is therefore $\Gamma = \frac{1}{2}LV$.

Shortly after leaving the tube the cylindrical vortex rolls up to form a vortex ring of circulation Γ which translates by self-induction (as determined by the Biot–Savart law (4.3.1)) at speed estimated by Kelvin to equal

$$V_t \sim \frac{\Gamma}{4\pi R_0}\left[\ln\left(\frac{8R_0}{\sigma}\right) - \frac{1}{4}\right] \approx \frac{VL}{8\pi R_0}\left[\ln\left(\frac{8R_0}{\sigma}\right) - \frac{1}{4}\right],$$

where $R_0 \sim 1.2R$ is the radius of the vortex ring and $\sigma \sim 0.2R_0$ is the radius of its core (assumed to be of circular cross section). The ring arrives at the flame after a time $t_\ell \sim \ell/V_t$.

The air on the axis of the vortex ring at its center forms a localized jet with velocity on the centerline equal to

$$V_j \sim \frac{VL}{4R_0}.$$

If the flame is extinguished it is because the vortex jet blows away the hot combusting gases from newly vaporized wax.

According to this sequence of events, the candle is only blown out because of the presence of the vortex. In its absence the flame would barely flicker under the influence of blowing by the potential velocity field V_φ. The following numerical estimates confirm this conclusion. Take $V = 10$ m/s, $L = 1$ cm, $R = 0.5$ cm, and $\ell = 0.3$ m. Then, $R_0 \approx 0.6$ cm, $\sigma \approx 0.12$ cm, $V_\varphi \approx 0.0007$ m/s, $V_t \approx 2.4$ m/s, $V_j \approx 4.2$ m/s, and $t_\ell \approx 0.13$ s.

4.4 Surface Force in Incompressible Flow Expressed in Terms of Vorticity

We now derive the following formula for the force $\mathbf{F}$ exerted on an incompressible fluid by a rigid body with surface S whose centre of volume has velocity $\mathbf{U}_0$:

$$\mathbf{F} + m_0 \frac{d\mathbf{U}_0}{dt} = \rho_0 \frac{d\mathbf{I}}{dt} \equiv \frac{\rho_0}{2} \frac{d}{dt} \int \mathbf{x} \wedge \boldsymbol{\omega}(\mathbf{x}, t) \, d^3\mathbf{x}, \qquad (4.4.1)$$

where $\mathbf{I}$ is the impulse defined by the integral in (4.3.3), including any contributions from bound vorticity within S, and m_0 is the mass of fluid displaced by the body.

With reference to Fig. 4.1.1, let

$\mathcal{F}$ = external force applied to the body to maintain its motion,
m = mass of the body.

The fluid is assumed to be at rest at infinity. Let V denote the fluid between a large closed surface Σ containing all of the vorticity and the surface S of the body, and let V_+ denote the interior of Σ including the volume Δ of the body. The center of volume of the body is assumed to be in motion at velocity $\mathbf{U}_0(t)$, but in general the body may also be rotating at some time dependent angular velocity $\boldsymbol{\Omega}$. The global equations of motion are

$$m\frac{d\mathbf{U}}{dt} + \rho_0 \frac{d}{dt} \int_V \mathbf{v}(\mathbf{x}, t) \, d^3\mathbf{x} = \mathcal{F} + \oint_\Sigma p(\mathbf{x}, t) \, d\mathbf{S}$$

$$m\frac{d\mathbf{U}}{dt} = \mathcal{F} - \mathbf{F}$$

where $\mathbf{U}$ is the velocity of the centre of mass of the body. Subtracting these equations and extending the volume integral to include the volume Δ occupied by the rigid body (where $\int_\Delta \mathbf{v} \, d^3\mathbf{x} = \Delta\mathbf{U}_0$), we can also write

$$\mathbf{F} + m_0 \frac{d\mathbf{U}_0}{dt} = \rho_0 \frac{d}{dt} \int_{V_+} \mathbf{v}(\mathbf{x}, t) \, d^3\mathbf{x} - \oint_\Sigma p(\mathbf{x}, t) \, d\mathbf{S}. \qquad (4.4.2)$$

Now let $\mathbf{v}(\mathbf{x}, t) = \operatorname{curl} \mathbf{A}$, where the vector potential $\mathbf{A}(\mathbf{x}, t)$ is defined by the Biot–Savart integral (4.3.1) taken over V_+, which includes the region occupied by the body. Then, (4.3.3) implies that $\mathbf{A} = \operatorname{curl}(\mathbf{I}(t)/4\pi|\mathbf{x}|)$ on Σ, so that

$$\int_{V_+} \mathbf{v}(\mathbf{x}, t) \, d^3\mathbf{x} = \int_{V_+} \operatorname{curl} \mathbf{A} \, d^3\mathbf{x} = -\int_\Sigma \mathbf{n} \wedge \mathbf{A} \, dS \to -\int_\Sigma \mathbf{n} \wedge \operatorname{curl}\left(\frac{\mathbf{I}(t)}{4\pi|\mathbf{x}|}\right) dS.$$

Similarly, $p \to -\rho_0 \partial\varphi/\partial t$ on Σ, where $\varphi = \mathrm{div}(\mathbf{I}(t)/4\pi\,|\mathbf{x}|)$ (see (4.3.3)). Hence,

$$-\oint_\Sigma p(\mathbf{x}, t)\, d\mathbf{S} \to \rho_0 \frac{d}{dt} \oint_\Sigma \varphi \mathbf{n}\, dS = \rho_0 \frac{d}{dt} \oint_\Sigma \mathbf{n}\, \mathrm{div}\left(\frac{\mathbf{I}(t)}{4\pi\,|\mathbf{x}|}\right) dS.$$

The right-hand side of (4.4.2) can therefore be written

$$\rho_0 \frac{d}{dt} \oint_\Sigma \left\{ -\mathbf{n} \wedge \mathrm{curl}\left(\frac{\mathbf{I}(t)}{4\pi\,|\mathbf{x}|}\right) + \mathbf{n}\, \mathrm{div}\left(\frac{\mathbf{I}(t)}{4\pi\,|\mathbf{x}|}\right) \right\} dS.$$

By the divergence theorem, the integral in this expression over the large, but arbitrary surface Σ can be replaced by an integration over the surface of a large sphere $|\mathbf{x}| = R$, because

$$\{\mathrm{curl\,curl} - \nabla\mathrm{div}\}(\mathbf{I}(t)/4\pi\,|\mathbf{x}|) = -\nabla^2(\mathbf{I}(t)/4\pi\,|\mathbf{x}|) \equiv 0 \quad \text{for } |\mathbf{x}| > 0.$$

On the sphere $\mathbf{n} = -\mathbf{x}/|\mathbf{x}|$, and the integrand equals $\mathbf{I}(t)/4\pi R^2$; the integral is therefore just equal to $\mathbf{I}(t)$. Thus, (4.4.2) reduces to the desired representation (4.4.1).

4.4.1 Bound Vorticity and the Added Mass

Consider the particular case of a rigid body accelerating without rotation at velocity $\mathbf{U}(t)$ in an otherwise unbounded, ideal incompressible fluid in the absence of vorticity (Fig. 4.4.1). We have seen previously (Section 3.8) that force exerted on the fluid can be written

$$F_i = M_{ij} \frac{dU_j}{dt},$$

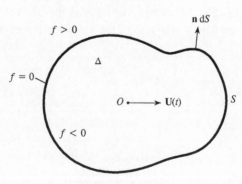

Fig. 4.4.1.

where M_{ij} is the added mass tensor (3.8.6). This result will now be derived from the integral formula (4.4.1).

There is no vorticity in the fluid, but the slipping of the ideal flow over S generates a singular distribution (a vortex sheet) of *bound vorticity* that must be used to evaluate the integral. To calculate the bound vorticity we need an expression for the velocity *everywhere* in space, including the region occupied by the body, where $\mathbf{v} \equiv \mathbf{U}$.

In the fluid, we can take

$$\mathbf{v} = U_j \nabla \varphi_j^*.$$

Therefore, by introducing a control surface $f(\mathbf{x}, t) = 0$ that coincides with the surface S of the body, with $f > 0$ in the fluid and $f < 0$ within S, the required formula for the velocity is

$$\mathbf{v} = U_j H(f) \nabla \varphi_j^* + U_j H(-f) \nabla x_j,$$

and the vorticity is

$$
\begin{aligned}
\boldsymbol{\omega} &= \operatorname{curl}\{U_j H(f) \nabla \varphi_j^* + U_j H(-f) \nabla x_j\} \\
&= -U_j \nabla H \wedge \nabla(x_j - \varphi_j^*) \quad (\text{where } \nabla H = \nabla H(f) = -\nabla H(-f)) \\
&\equiv \operatorname{curl}\{U_j(x_j - \varphi_j^*) \nabla H\}.
\end{aligned}
$$

Equation (4.4.1) accordingly gives the force in the form

$$\mathbf{F} = -m_0 \frac{d\mathbf{U}}{dt} + \frac{\rho_0}{2} \frac{d}{dt} \int \mathbf{x} \wedge \operatorname{curl}\{U_j(x_j - \varphi_j^*) \nabla H\} \, d^3\mathbf{x}.$$

Now the vector $\mathbf{A} = U_j(x_j - \varphi_j^*) \nabla H$ vanishes except on the surface S of the body. The identities

$$\mathbf{x} \wedge \operatorname{curl} \mathbf{A} = 2\mathbf{A} + \nabla(\mathbf{x} \cdot \mathbf{A}) - \frac{\partial}{\partial x_j}(x_j \mathbf{A}), \qquad \int (\cdot) \nabla H \, d^3\mathbf{x} = \oint_S (\cdot) \, d\mathbf{S} \tag{4.4.3}$$

therefore imply that

$$
\begin{aligned}
F_i &= -m_0 \frac{dU_i}{dt} + \rho_0 \frac{d}{dt} \int U_j(x_j - \varphi_j^*) \frac{\partial H}{\partial x_i} \, d^3\mathbf{x} \\
&= -m_0 \frac{dU_i}{dt} + \rho_0 \frac{d}{dt} \oint_S U_j(x_j - \varphi_j^*) n_i \, dS.
\end{aligned}
$$

But

$$\rho_0 \oint_S x_j n_i \, dS = \rho_0 \Delta \delta_{ij} \equiv m_0 \delta_{ij},$$

where Δ = volume contained by S, and

$$\rho_0 \oint_S -\varphi_j^* n_i \, dS = M_{ij},$$

where M_{ij} is the added mass tensor (3.8.8). Therefore, force on fluid in irrotational flow $\equiv F_i = M_{ij} \frac{dU_j}{dt}$.

4.4.2 Force Exerted on an Incompressible Fluid by a Moving Body

The integral in (4.4.1) defining the value of $d\mathbf{I}/dt$ can be transformed to remove the strong dependence of the integrand on the bound vorticity on S. This vorticity is produced both by motion of S and by relative motion between S and the fluid induced by free vorticity in the flow. Thus, any attempt to recast $d\mathbf{I}/dt$ must be strongly influenced by both the shape and motion of S. We consider only the important special case of a body in *translational motion without rotation* at velocity $\mathbf{U}(t)$, and show that the ith component of the force $\mathbf{F}$ exerted on the fluid can also be written

$$F_i = M_{ij} \frac{dU_j}{dt} - \rho_0 \int_V \nabla X_i \cdot \boldsymbol{\omega} \wedge \mathbf{v}_{\text{rel}} \, d^3\mathbf{x} - \eta \oint_S \nabla X_i \cdot \boldsymbol{\omega} \wedge d\mathbf{S}, \quad \mathbf{v}_{\text{rel}} = \mathbf{v} - \mathbf{U},$$

$$(4.4.4)$$

where $\mathbf{v}_{\text{rel}}$ is the fluid velocity relative to the translational velocity of S, $X_i = x_i - \varphi_i^*(\mathbf{x}, t)$ is the Kirchhoff vector already encountered in the definition of the compact Green's function (Section 3.4), and $M_{ij} = M_{ji} = -\rho_0 \oint_S n_j \varphi_i^* \, dS$ is the added mass tensor. X_i represents the velocity potential of an ideal flow past S that has unit speed in the i direction at large distances from S (it depends on t because a fixed coordinate system is being used).

The first term on the right of (4.4.4) represents the *inviscid* component of the force, associated with the added mass. The contribution from free vorticity is furnished by the volume integral; the final term arises from frictional effects on S, which are relatively small at large Reynolds numbers. Now $\mathbf{v}_{\text{rel}} = 0$ on S, and therefore the contribution to the volume integral from vorticity close to and on S is negligible; indeed, even in the inviscid limit there is no contribution to the integral from the surface vortex sheet forming the bound vorticity, because ∇X_i and the relative Lamb vector $\boldsymbol{\omega} \wedge \mathbf{v}_{\text{rel}}$ are orthogonal on S.

To derive this formula from (4.4.1) we introduce the usual control surface $f(\mathbf{x}, t) = 0$ enclosing S, with $f > 0$ in the outer fluid region, multiply Crocco's homentropic momentum equation (4.2.5) by $H \equiv H(f)$, and take the curl of the resulting equation. Using the formula

$$\frac{DH}{Dt} \equiv \frac{\partial H}{\partial t} + \mathbf{v} \cdot \nabla H = 0,$$

and the no-slip condition on S, we find

$$\frac{\partial}{\partial t}(H\omega) = -\frac{\partial}{\partial t}(\nabla H \wedge \mathbf{U}) - \mathrm{curl}((\nabla H \cdot \mathbf{U})\mathbf{U}) - \nabla H \wedge \nabla B$$
$$- \mathrm{curl}(H\omega \wedge \mathbf{v}) - \nu\,\mathrm{curl}(H\,\mathrm{curl}\,\omega).$$

Then, because $\omega = \mathbf{0}$ within S,

$$\frac{d}{dt}\int \mathbf{x} \wedge \omega(\mathbf{x}, t)\, d^3\mathbf{x} = \frac{d}{dt}\int \mathbf{x} \wedge (H\omega)\, d^3\mathbf{x} = \int \mathbf{x} \wedge \frac{\partial}{\partial t}(H\omega)\, d^3\mathbf{x}$$

$$= -\int \mathbf{x} \wedge \frac{\partial}{\partial t}(\nabla H \wedge \mathbf{U})\, d^3\mathbf{x} - \int \mathbf{x} \wedge \mathrm{curl}\,((\nabla H \cdot \mathbf{U})\mathbf{U})\, d^3\mathbf{x}$$

$$- \int \mathbf{x} \wedge (\nabla H \wedge \nabla B)\, d^3\mathbf{x} - \int \mathbf{x} \wedge \mathrm{curl}\,(H\omega \wedge \mathbf{v})\, d^3\mathbf{x}$$

$$- \nu \int \mathbf{x} \wedge \mathrm{curl}\,(H\,\mathrm{curl}\,\omega)\, d^3\mathbf{x}$$

$$= 2\Delta\frac{d\mathbf{U}}{dt} + 0 + 2\oint_S B\, d\mathbf{S} - 2\int_V \omega \wedge \mathbf{v}\, d^3\mathbf{x} - 2\nu\oint_S \omega \wedge d\mathbf{S},$$

where the last line follows by use of the identities (4.4.3). Thus, adopting suffix notation,

$$\frac{dI_i}{dt} \equiv \frac{1}{2}\frac{d}{dt}\int (\mathbf{x} \wedge \omega)_i\, d^3\mathbf{x} = \Delta\frac{dU_i}{dt} + \oint_S Bn_i\, dS - \int_V \nabla x_i \cdot (\omega \wedge \mathbf{v})\, d^3\mathbf{x}$$
$$- \nu\oint_S \nabla x_i \cdot \omega \wedge d\mathbf{S}. \qquad (4.4.5)$$

The surface integral $\oint_S Bn_i\, dS$ can be eliminated by recalling that

$$\nabla^2\varphi_i^* = 0, \qquad \frac{\partial\varphi_i^*}{\partial x_{\mathrm{n}}} \equiv n_j\frac{\partial\varphi_i^*}{\partial x_j} = n_i \quad \text{on } S.$$

Then, because $\nabla\varphi_i^* \sim O(1/|\mathbf{x}|^3)$ as $|\mathbf{x}| \to \infty$, the divergence theorem shows that $\oint_S Bn_i\, dS = -\int_V \mathrm{div}(\nabla\varphi_i^* B)\, d^3\mathbf{x} \equiv -\int_V \nabla\varphi_i^* \cdot \nabla B\, d^3\mathbf{x}$. Hence, using Crocco's equation (4.2.5)

$$\oint_S Bn_i\, dS = \int_V \mathrm{div}\left(\varphi_i^*\frac{\partial\mathbf{v}}{\partial t}\right) d^3\mathbf{x} + \int_V \nabla\varphi_i^* \cdot \omega \wedge \mathbf{v}\, d^3\mathbf{x}$$
$$- \nu\int_V \mathrm{div}(\nabla\varphi_i^* \wedge \omega)\, d^3\mathbf{x}.$$

The first and last integrals on the right are transformed further by the divergence theorem, for the first

$$\int_V \operatorname{div}\left(\varphi_i^* \frac{\partial \mathbf{v}}{\partial t}\right) d^3\mathbf{x} = \frac{M_{ij}}{\rho_0}\frac{dU_j}{dt},$$

where $M_{ij} = M_{ji} = -\rho_0 \oint_S n_j \varphi_i^* \, dS$ is the added mass coefficient of (3.8.6). For the last

$$-\nu \int_V \operatorname{div}(\nabla \varphi_i^* \wedge \boldsymbol{\omega}) \, d^3\mathbf{x} = \nu \oint_S \nabla \varphi_i^* \cdot \boldsymbol{\omega} \wedge d\mathbf{S}.$$

Thus, substituting for $\oint_S B n_i \, dS$ in (4.4.5), we find

$$\frac{dI_i}{dt} = \Delta \frac{dU_i}{dt} + \frac{M_{ij}}{\rho_0}\frac{dU_j}{dt} - \int_V \nabla X_i \cdot \boldsymbol{\omega} \wedge \mathbf{v} \, d^3\mathbf{x} - \nu \oint_S \nabla X_i \cdot \boldsymbol{\omega} \wedge d\mathbf{S}. \quad (4.4.6)$$

But the identity

$$\nabla X_i \cdot \boldsymbol{\omega} \wedge \mathbf{U} = \operatorname{div}(\mathbf{U}(\mathbf{v}\cdot\nabla X_i) - \mathbf{v}(\mathbf{U}\cdot\nabla X_i) - (\mathbf{v}\cdot\mathbf{U})\nabla X_i)$$

implies that $\int_V \nabla X_i \cdot \boldsymbol{\omega} \wedge \mathbf{U} \, d^3\mathbf{x} = 0$, so that (4.4.6) can also be written

$$\frac{dI_i}{dt} = \Delta \frac{dU_i}{dt} + \frac{M_{ij}}{\rho_0}\frac{dU_j}{dt} - \int_V \nabla X_i \cdot \boldsymbol{\omega} \wedge \mathbf{v}_{\mathrm{rel}} \, d^3\mathbf{x} - \nu \oint_S \nabla X_i \cdot \boldsymbol{\omega} \wedge d\mathbf{S} \quad (4.4.7)$$

where $\mathbf{v}_{\mathrm{rel}} = \mathbf{v} - \mathbf{U}$.

Equation (4.4.4) is now obtained by substituting from (4.4.7) into (4.4.1) (recalling that $m_0 = \rho_0\Delta$).

4.4.3 Stokes Drag on a Sphere

The first term on the right-hand side of (4.4.4) is the force necessary to accelerate the added mass of the body. The ith component of the viscous skin friction is $-\eta \oint_S (\boldsymbol{\omega} \wedge d\mathbf{S})_i \equiv -\eta \oint_S \nabla x_i \cdot \boldsymbol{\omega} \wedge d\mathbf{S}$. Thus, (because $X_i = x_i - \varphi_i^*$) the net contribution of the normal pressure forces on S is represented in (4.4.4) by the terms

$$-\rho_0 \int_V \nabla X_i \cdot \boldsymbol{\omega} \wedge \mathbf{v}_{\mathrm{rel}} \, d^3\mathbf{x} + \eta \oint_S \nabla \varphi_i^* \cdot \boldsymbol{\omega} \wedge d\mathbf{S}.$$

The second, viscous component is comparable in magnitude to the skin friction, and is produced by the pressure field established by the surface shear stress.

The necessity for such a term is vividly illustrated by the Stokes drag on a sphere. Let the sphere have radius a and translate at constant velocity $\mathbf{U} = (U, 0, 0)$, $U > 0$, along the x_1 axis. At very small Reynolds numbers $\mathrm{Re} = aU/\nu \ll 1$

the inertial terms $\boldsymbol{\omega} \wedge \mathbf{v}$ and $\nabla(\frac{1}{2}v^2)$ can be discarded from Crocco's equation (4.2.5), which (for incompressible flow) reduces to the creeping flow equation

$$\nabla p = -\eta \operatorname{curl} \boldsymbol{\omega} \tag{4.4.8}$$

in a reference frame moving with the sphere. Both the pressure and the vorticity therefore satisfy Laplace's equation $\nabla^2 p = 0$, $\nabla^2 \boldsymbol{\omega} = \mathbf{0}$. By symmetry p must vary linearly with η and $\mathbf{U} \cdot \mathbf{x}$, and the condition that p should vanish at large distances from the sphere supplies the dipole solution

$$p = C\eta \frac{\mathbf{U} \cdot \mathbf{x}}{|\mathbf{x}|^3} \equiv -C\eta \operatorname{div}\left(\frac{\mathbf{U}}{|\mathbf{x}|}\right), \quad |\mathbf{x}| > a,$$

where C is a constant.

Similarly, $\boldsymbol{\omega}$ must be a linear function of $\mathbf{U} \wedge \mathbf{x}$: the identity $\operatorname{curl} \operatorname{curl}(\mathbf{U}/|\mathbf{x}|) = \operatorname{grad} \operatorname{div}(\mathbf{U}/|\mathbf{x}|)$ and Equation (4.4.8) imply that $\boldsymbol{\omega} = C \operatorname{curl}(\mathbf{U}/|\mathbf{x}|) \equiv C(\mathbf{U} \wedge \mathbf{x})/|\mathbf{x}|^3$ ($|\mathbf{x}| > a$). The value of C is most easily found by substituting this expression for $\boldsymbol{\omega}$ into the Biot–Savart formula (4.3.1) and evaluating the right-hand side at the centre $\mathbf{x} = \mathbf{0}$ of the sphere, where $\mathbf{v} \equiv \mathbf{U}$. This yields $C = 3a/2$, and therefore

$$\boldsymbol{\omega} = \operatorname{curl}\left(\frac{3a\mathbf{U}}{2|\mathbf{x}|}\right), \quad |\mathbf{x}| > a. \tag{4.4.9}$$

The net force F_1 on the fluid is in the x_1 direction, and is given by the final integral on the right of (4.4.4). It is equal in magnitude to $D_s + D_p$, where D_s and D_p are the respective components of the Stokes drag on the sphere produced by the skin friction and the viscous surface pressure. For the sphere $\varphi_1^* = -a^3 x_1/2|\mathbf{x}|^3$, and we readily calculate $F_1 = 6\pi \eta U a$, and

$$D_s = \eta \oint_S (\boldsymbol{\omega} \wedge d\mathbf{S})_1 = -4\pi \eta U a, \qquad D_p = -\eta \oint_S \nabla \varphi_1^* \cdot \boldsymbol{\omega} \wedge d\mathbf{S} = -2\pi \eta U a.$$

The pressure drag is therefore equal to half the skin-friction drag.

This interpretation of D_p as the component of drag attributable to the normal pressure forces on the sphere can be confirmed directly using the creeping flow approximation (4.4.8). Because $n_1 = \mathbf{n} \cdot \nabla \varphi_1^*$ on S we find, using the divergence theorem,

$$D_p = -\oint_S p n_1 \, dS \equiv -\oint_S p \nabla \varphi_1^* \cdot d\mathbf{S} = \int_V \nabla p \cdot \nabla \varphi_1^* \, d^3\mathbf{x}$$

$$= -\eta \int_V \operatorname{curl} \boldsymbol{\omega} \cdot \nabla \varphi_1^* \, d^3\mathbf{x} = -\eta \oint_S \nabla \varphi_1^* \cdot \boldsymbol{\omega} \wedge d\mathbf{S},$$

where (4.4.8) and the identity $\operatorname{div}(\mathbf{A} \wedge \mathbf{B}) = \operatorname{curl} \mathbf{A} \cdot \mathbf{B} - \mathbf{A} \cdot \operatorname{curl} \mathbf{B}$ have been used on the second line.

4.5 The Complex Potential

The remainder of this chapter is devoted to a brief outline of the complex potential representation of two-dimensional, incompressible flows and its application to determine the equation of motion of a line vortex in such flows. The results will be applied in later chapters to investigate simple models of sound production by vortices interacting with surfaces.

4.5.1 Laplace's Equation in Two Dimensions

Suppose that

$$w(z) = \varphi(x, y) + i\psi(x, y), \qquad z = x + iy$$

is *regular* (analytic) in a region $\mathcal{D}$ of the z-plane. The real and imaginary parts $\varphi(x, y)$ and $\psi(x, y)$ are solutions of Laplace's equation.

$$\frac{\partial^2 \varphi}{\partial x^2} + \frac{\partial^2 \varphi}{\partial y^2} = 0, \qquad \frac{\partial^2 \psi}{\partial x^2} + \frac{\partial^2 \psi}{\partial y^2} = 0 \quad \text{in } \mathcal{D}.$$

Let $f(z)$ be regular in $\mathcal{D}$, and define a (conformal) transformation $Z = f(z)$ of $\mathcal{D}$ into a region $\mathcal{D}'$ in the plane of $Z = X + iY$. Let $\mathcal{W}(Z)$ be regular in $\mathcal{D}'$ with real and imaginary parts $\Phi(X, Y)$, $\Psi(X, Y)$. Then,

$$\frac{\partial^2 \Phi}{\partial X^2} + \frac{\partial^2 \Phi}{\partial Y^2} = 0, \qquad \frac{\partial^2 \Psi}{\partial X^2} + \frac{\partial^2 \Psi}{\partial Y^2} = 0 \quad \text{in } \mathcal{D}'.$$

The transformation $Z = f(z)$ permits us to define a corresponding function $w(z) \equiv \varphi(x, y) + i\psi(x, y) = \mathcal{W}(f(z))$, which is regular in $\mathcal{D}$, with derivative $w'(z) = f'(z)\mathcal{W}'(f(z))$. For corresponding points in $\mathcal{D}$ and $\mathcal{D}'$ we have

$$\varphi(x, y) = \Phi(X(x, y), Y(x, y)), \qquad \psi(x, y) = \Psi(X(x, y), Y(x, y)).$$

In other words, the solutions Φ and Ψ of Laplace's equation in $\mathcal{D}'$ are also solutions of Laplace's equation in $\mathcal{D}$.

These results have the following significance. The solution of Laplace's equation within a given two-dimensional bounded region $\mathcal{D}$ is equivalent to the solution of Laplace's equation within the transformed region $\mathcal{D}'$. If it is possible to solve the latter problem, the solution to the original problem in $\mathcal{D}$ can be found by transforming back to the z plane. Difficulties may arise at isolated points where $f'(z) = 0$ and at points where $f(z)$ ceases to be regular, but these can usually be dealt with by careful examination of the behavior of the transformation near such points.

4.5.2 Hydrodynamics in Two Dimensions

Irrotational motion of an ideal, incompressible fluid in planes parallel to the xy plane can be investigated by introducing the *complex potential* $w(z) = \varphi(x, y) + i\psi(x, y)$. The velocity

$$\mathbf{v} = \nabla\varphi = (\partial\varphi/\partial x, \partial\varphi/\partial y).$$

The function ψ is called the *stream function*. For *steady* motion the velocity at (x, y) does not change with time, and the fluid particles travel along a fixed system of streamlines each of which is a member of the family of curves $\psi(x, y) =$ constant.

Both $\varphi(x, y)$ and $\psi(x, y)$ are solutions of Laplace's equation and satisfy the Cauchy–Riemann equations:

$$\frac{\partial\varphi}{\partial x} = \frac{\partial\psi}{\partial y}, \qquad \frac{\partial\varphi}{\partial y} = -\frac{\partial\psi}{\partial x},$$

which imply that $\nabla\varphi \cdot \nabla\psi = 0$, i.e., that the streamlines intersect the equipotentials $\varphi =$ constant at right angles. If $\mathbf{v} = (u, v)$, then the *complex velocity*

$$w'(z) = \frac{\partial\varphi}{\partial x} - i\frac{\partial\varphi}{\partial y} \equiv u - iv$$

is also regular.

The fact that $w(z)$ is a regular function of z can greatly simplify the solution of many problems. This will be illustrated by consideration of two methods based on the theory of complex variables.

Method 1 The real and imaginary parts of every regular function $w(z)$ determine the velocity potential and stream function of a possible flow. A catalog of flows can therefore be constructed by studying the properties of arbitrarily selected $w(z)$.

Example 1 $w = Uz$, $U =$ real constant:

$$\varphi = Ux, \qquad \psi = Uy. \qquad \text{Thus, } \mathbf{v} = (U, 0).$$

The motion is uniform at speed U along streamlines parallel to the x direction.

Example 2

$$w = U\left(z + \frac{a^2}{z}\right), \quad U = \text{real constant}, a > 0, \quad |z| > a. \quad (4.5.1)$$

At large distances from the origin $w \to Uz$, and the motion becomes uniform at speed U parallel to the x axis. In terms of the polar form $z = re^{i\theta}$,

$$w = U\left(re^{i\theta} + \frac{a^2}{r}e^{-i\theta}\right). \quad \text{Thus, } \varphi = U\cos\theta\left(r + \frac{a^2}{r}\right).$$

The radial component of velocity

$$\frac{\partial\varphi}{\partial r} = U\cos\theta\left(1 - \frac{a^2}{r^2}\right)$$

vanishes at $r = a$. The motion therefore represents steady flow in the x direction past a rigid cylinder of radius a with centre at the origin (Fig. 4.5.1; c.f., Section 3.6).

Example 3

$$w = -iU\left(z - \frac{a^2}{z}\right), \quad U = \text{real constant}, \quad |z| > a > 0,$$

$$\left(\varphi = U\sin\theta\left(r + \frac{a^2}{r}\right)\right), \quad (4.5.2)$$

describes potential flow in the y direction past a rigid cylinder of radius a with center at the origin.

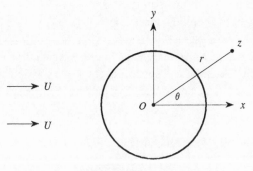

Fig. 4.5.1.

Example 4 The function

$$w = \frac{1}{2\pi} \ln z, \quad \left(\varphi = \frac{1}{2\pi} \ln r, \psi = \frac{\theta}{2\pi}, z = re^{i\theta} \right),$$

is regular except at $z = 0$. The flow is radially outward from the origin along streamlines $\theta = $ constant, at speed $\partial\varphi/\partial r = 1/2\pi r$. The origin is a singularity of the flow where fluid is *created* at a rate equal to $\oint_C \nabla\varphi \cdot \mathbf{n}\, ds$, where C is any simple closed curve enclosing the origin with outward normal $\mathbf{n}$, and ds is the element of arc length on C. In particular, taking C to be a circle of radius r,

$$\oint_C \nabla\varphi \cdot \mathbf{n}\, ds = \int_0^{2\pi} \frac{\partial\varphi}{\partial r} r\, d\theta = 1.$$

The origin is therefore a simple source of *unit* strength. When the source is situated at $z_0 = x_0 + iy_0$

$$w = \frac{1}{2\pi} \ln(z - z_0), \quad \left(\varphi = \frac{1}{2\pi} \ln|z - z_0| = \frac{1}{2\pi} \ln\sqrt{(x - x_0)^2 + (y - y_0)^2} \right).$$

Example 5 The function

$$w = \frac{-i\Gamma}{2\pi} \ln z, \quad \left(\varphi = \frac{\Gamma\theta}{2\pi}, \psi = -\frac{\Gamma}{2\pi} \ln r, z = re^{i\theta} \right),$$

is regular except at $z = 0$, and describes the irrotational flow outside a *line vortex* of strength Γ concentrated at $z = 0$. The streamlines are circles centered at $z = 0$, and the flow speed is $\partial\varphi/r\partial\theta = \Gamma/2\pi r$ in the anticlockwise direction (for $\Gamma > 0$). The circulation $\int_C \mathbf{v} \cdot d\mathbf{x} = \Gamma$, where C is any contour encircling the vortex once, and the contour is traversed in the *positive* direction (with the *interior on the left*). When the vortex is at $z_0 = x_0 + iy_0$

$$w = \frac{-i\Gamma}{2\pi} \ln(z - z_0).$$

Example 6 The function

$$w = \frac{1}{2\pi}(\ln(z - z_0) + \ln(z - z_0^*)), \quad \left(\varphi = \frac{1}{2\pi}(\ln r_1 + \ln r_2) \right),$$

represents the flow produced by two unit point sources located at $z_0 = x_0 + iy_0$ and $z_0^* = x_0 - iy_0$ (Fig. 4.5.2). The motion is symmetric with respect to the x axis, and $\partial\varphi/\partial y = 0$ on $y = 0$. Therefore, in the region $y > 0$ the potential also describes the flow produced by a point source at z_0 *adjacent to a rigid wall* at $y = 0$ (the presence of the wall is said to be accounted for by an image source).

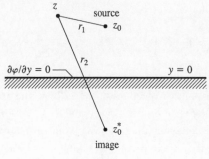

Fig. 4.5.2.

Example 7 The function

$$w = \frac{-i\Gamma}{2\pi} \ln(z - z_0) + \frac{i\Gamma}{2\pi} \ln(z - z_0^*),$$

represents the flow produced by two line vortices of circulations $\pm\Gamma$ respectively at $z_0 = x_0 + iy_0$, $z_0^* = x_0 - iy_0$ (Fig. 4.5.3). The stream function $\psi = \text{Im } w = 0$ on the x axis, which is therefore a streamline of the flow, on which $\partial\varphi/\partial y = 0$. In the region $y > 0$ the potential describes the flow produced by a vortex of strength Γ at z_0 *adjacent to a rigid wall* at $y = 0$ (which is accounted for by an equal and opposite image vortex). Each vortex translates parallel to the wall at speed $u = \Gamma/4\pi y_0$ determined by the velocity potential of its image. The mean value of the local rotational flow produced by the self-potential of each vortex (Example 5) vanishes on the vortex axis, and cannot therefore cause it to translate.

Method 2 The flow past a system of rigid boundaries in the z plane is represented by means of a conformal transformation $Z = f(z)$ by an equivalent flow in the Z plane. The transformation is usually chosen to simplify the boundary

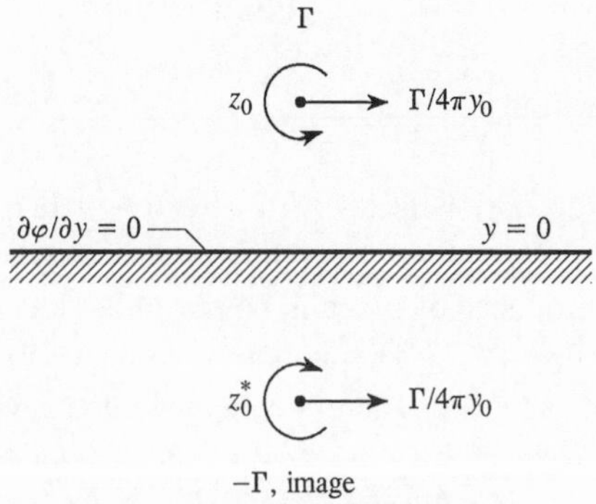

Fig. 4.5.3.

conditions, thereby permitting the solution in the Z plane to be found in a relatively straightforward manner. Point source and vortex singularities of the flow are *preserved* under the transformation. Indeed, if $Z = Z_0$ is the image of a vortex of strength Γ at $z = z_0$, the complex potential in the neighborhood of Z_0 (where $Z - Z_0 \approx f'(z_0)(z - z_0)$) is determined by

$$\mathcal{W}(Z) = w(z) = \frac{-i\Gamma}{2\pi} \ln(z - z_0) + \text{terms finite at } z_0$$

$$= \frac{-i\Gamma}{2\pi} \ln\left(\frac{Z - Z_0}{f'(z_0)}\right) + \text{terms finite at } Z_0$$

$$= \frac{-i\Gamma}{2\pi} \ln(Z - Z_0) + \text{terms finite at } Z_0.$$

The vortex in the z plane therefore maps into an equal vortex at the image point in the Z plane.

Example 8 Derive the following formula for the velocity potential of irrotational flow around the edge of the rigid half-plane $x < 0$, $y = 0$ in terms of polar coordinates (r, θ):

$$\varphi = \alpha\sqrt{r}\sin\frac{\theta}{2}, \quad \alpha = \text{a real constant,}$$

and plot the streamlines.

The transformation $Z = i\sqrt{z}$ maps the z plane cut along the negative real axis (so that $-\pi < \arg z < \pi$) onto the upper half of the Z plane. The complex potential of flow in the positive X direction parallel to the boundary $Y = 0$ in the Z plane corresponds to flow around the edge of the half-plane in the clockwise sense, and has the general representation $W = UZ$, where U is real. In the z plane this becomes

$$w = iU\sqrt{z} \equiv -U\sqrt{r}\sin\left(\frac{\theta}{2}\right) + iU\sqrt{r}\cos\left(\frac{\theta}{2}\right), \quad -\pi < \theta < \pi.$$

The polar representation of the velocity is therefore

$$\mathbf{v} = (v_r, v_\theta) = \left(\frac{\partial\varphi}{\partial r}, \frac{1}{r}\frac{\partial\varphi}{\partial\theta}\right) = \frac{-U}{2\sqrt{r}}\left(\sin\frac{\theta}{2}, \cos\frac{\theta}{2}\right).$$

This satisfies the rigid wall condition on the half-plane because the component of velocity normal to the wall is v_θ, which vanishes at $\theta = \pm\pi$. The streamlines of the flow are the parabolas

$$\sqrt{r}\cos\left(\frac{\theta}{2}\right) = \text{constant}, \quad \text{i.e., } y = \pm 2\beta\sqrt{1 - \frac{x}{\beta}},$$

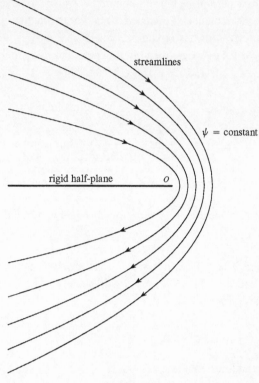

streamlines

ψ = constant

rigid half-plane o

Fig. 4.5.4.

where $x < \beta$, β being a positive constant, as shown in Fig. 4.5.4. When $U > 0$ fluid particles travel along the parabolic streamlines around the edge in the *clockwise* direction. The streamline for $\beta = 0$ corresponds to the upper and lower surfaces of the half-plane, which maps into the streamline $Y = 0$ on the surface of the wall in the Z plane. The flow velocity becomes infinite like $1/\sqrt{r}$ as $r \to 0$ at the sharp edge.

4.6 Motion of a Line Vortex

In two-dimensional incompressible, inviscid flow in planes parallel to $x_3 = 0$ the vortex lines are all parallel to the x_3 direction, and the vorticity equation (4.2.7) reduces to

$$\frac{D\omega_3}{Dt} = 0.$$

A line vortex is therefore convected without change at the local velocity at its core. For a vortex of strength Γ at $z = z_0(t)$ in the plane of $z = x_1 + ix_2$, the velocity becomes infinite as the core is approached because of the singular velocity induced by its self-potential

$$-\frac{i\Gamma}{2\pi} \ln(z - z_0).$$

But the rotational flow around the core induced by the vortex cannot induce motion in itself, and this potential must be removed from the complex potential $w(z)$ before calculating the convection velocity of the vortex.

In applications the complex potential $w(z)$ usually arises in the form

$$w(z) = -\frac{i\Gamma}{2\pi} \ln(\zeta(z) - \zeta(z_0)) + F(z), \tag{4.6.1}$$

where $\zeta(z)$, $F(z)$ are regular functions of z in the neighborhood of the vortex core at $z = z_0$. In particular, when $|z - z_0|$ is small we have

$$\zeta(z) = \zeta(z_0) + (z - z_0)\zeta'(z_0) + \frac{(z - z_0)^2}{2}\zeta''(z_0) + \cdots,$$

where the primes denote differentiation with respect to z. Thus, subtracting the self-potential from $w(z)$ we find, near the vortex,

$$W(z) = w(z) + \frac{i\Gamma}{2\pi} \ln(z - z_0)$$

$$= -\frac{i\Gamma}{2\pi} \ln(\zeta(z) - \zeta(z_0)) + \frac{i\Gamma}{2\pi} \ln(z - z_0) + F(z)$$

$$\approx -\frac{i\Gamma}{2\pi} \ln\left[\zeta'(z_0) + \frac{1}{2}\zeta''(z_0)(z - z_0)\right] + F(z). \tag{4.6.2}$$

The complex velocity of the vortex is $W'(z_0) \equiv \{W'(z)\}_{z=z_0}$, i.e.,

$$\frac{dz_0^*}{dt} \equiv \frac{dx_{01}}{dt} - i\frac{dx_{02}}{dt} = W'(z_0).$$

Using (4.6.2) this becomes

$$\frac{dx_{01}}{dt} - i\frac{dx_{02}}{dt} = -\frac{i\Gamma\zeta''(z_0)}{4\pi\zeta'(z_0)} + F'(z_0). \tag{4.6.3}$$

The real and imaginary parts of this equation supply two nonlinear first-order ordinary differential equations for the position $(x_{01}(t), x_{02}(t))$ of the vortex at time t.

4.6.1 Numerical Integration of the Vortex Path Equation

In most cases it is necessary to integrate equation (4.6.3) numerically. The time and space variables should first be nondimensionalized with respect to convenient time and length scales defined by the problem (several examples are discussed in Chapter 8). The integration is started from a prescribed point on the trajectory through which the vortex is required to pass.

Let us consider integration by means of a fourth-order *Runge–Kutta* algorithm. Write the equation of motion (4.6.3) in the form

$$\frac{dz_0}{dt} = f^*(z_0), \quad \text{where } f(z_0) = -\frac{i\Gamma\zeta''(z_0)}{4\pi\zeta'(z_0)} + F'(z_0),$$

and let h be a suitably small integration time step (which need not be constant). Assume that at time t_n the vortex is at $z_0(t_n) = z_0^n$. To determine the complex position z_0^{n+1} at time $t_{n+1} = t_n + h$, we evaluate

$$k_1 = hf^*(z_0^n), \qquad k_2 = hf^*\left(z_0^n + \frac{1}{2}k_1\right), \qquad k_3 = hf^*\left(z_0^n + \frac{1}{2}k_2\right),$$

$$k_4 = hf^*(z_0^n + k_3),$$

and then find

$$z_0^{n+1} = z_0^n + \tfrac{1}{6}(k_1 + 2k_2 + 2k_3 + k_4).$$

Example 1 Calculate the trajectory of a line vortex of strength $\Gamma > 0$ adjacent to a rigid half-plane lying along the negative real axis ($x_1 < 0$, $x_2 = 0$; Fig. 4.6.1a).
  The transformation

$$\zeta = i\sqrt{z}, \qquad z = x_1 + ix_2, \qquad -\pi < \arg z < \pi, \tag{4.6.4}$$

maps the fluid region $-\pi < \arg z < \pi$ into the upper half $\text{Im}\,\zeta > 0$ of the ζ plane (Fig. 4.6.1b). Let the vortex at $z_0(t)$ map into a vortex at $\zeta = \zeta_0(t)$. The velocity potential $w(\zeta)$ of the motion in the ζ plane is found by introducing an image vortex of strength $-\Gamma$ at $\zeta = \zeta_0^*(t)$, as described in Example 7 of Section 4.5, in which case

$$w = \frac{-i\Gamma}{2\pi}\ln(\zeta - \zeta_0) + \frac{i\Gamma}{2\pi}\ln(\zeta - \zeta_0^*).$$

In the z plane this becomes

$$w(z) = -\frac{i\Gamma}{2\pi}\ln(\zeta(z) - \zeta(z_0)) + \frac{i\Gamma}{2\pi}\ln(\zeta(z) - \zeta^*(z_0)),$$

(a)

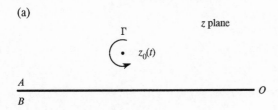

z plane

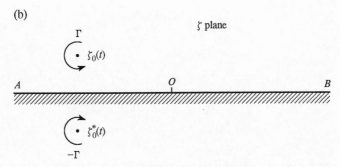

(b)

ζ plane

Fig. 4.6.1.

which is of the form (4.6.1). Hence, the equation of motion (4.6.3) becomes

$$\frac{dx_{01}}{dt} - i\frac{dx_{02}}{dt} = \frac{i\Gamma}{8\pi z_0} + \frac{i\Gamma}{4\pi\sqrt{z_0}[\sqrt{z_0} + (\sqrt{z_0})^*]}.$$

This can be integrated in closed form. Let $z_0 = re^{i\theta}$. Then the real and imaginary parts of the equation are

$$\frac{dx_{01}}{dt} \equiv \cos\theta\frac{dr}{dt} - r\sin\theta\frac{d\theta}{dt} = \frac{\Gamma}{8\pi r}\left(\sin\theta + \tan\frac{\theta}{2}\right),$$

$$\frac{dx_{02}}{dt} \equiv \sin\theta\frac{dr}{dt} + r\cos\theta\frac{d\theta}{dt} = -\frac{\Gamma}{8\pi r}(\cos\theta + 1).$$

Therefore,

$$\frac{dr}{dt} = -\frac{\Gamma}{8\pi r}\tan\frac{\theta}{2}, \qquad \frac{d\theta}{dt} = -\frac{\Gamma}{4\pi r^2} \tag{4.6.5}$$

that is,

$$r\frac{d\theta}{dr} = 2\cot\frac{1}{2}\theta$$

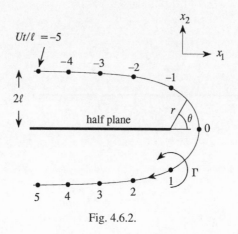

Fig. 4.6.2.

Thus,

$$r = \ell \sec \frac{1}{2}\theta, \qquad \ell = \text{constant.} \tag{4.6.6}$$

This is the polar equation of the trajectory plotted in Fig. 4.6.2. The constant length ℓ is equal to the distance of closest approach of the vortex to the edge of the half-plane, which occurs at $\theta = 0$. Substituting for r in the second of equations (4.6.5), we find

$$\sec^2\left(\frac{1}{2}\theta\right)\frac{d\theta}{dt} = -\frac{\Gamma}{4\pi\ell^2}. \quad \text{Thus, } \theta = 2\tan^{-1}\left(-\frac{\Gamma t}{8\pi\ell^2}\right),$$

where time is measured from the instant at which $\theta = 0$. The dependence of r on t is now obtained by substituting into (4.6.6).

Collecting together these results we have

$$r = \ell\sqrt{1 + \left(\frac{Ut}{\ell}\right)^2}, \qquad \theta = 2\tan^{-1}\left(-\frac{Ut}{\ell}\right);$$

$$\frac{x_{01}}{\ell} = \frac{1 - (Ut/\ell)^2}{\sqrt{1 + (Ut/\ell)^2}}, \qquad \frac{x_{02}}{\ell} = \frac{-2Ut/\ell}{\sqrt{1 + (Ut/\ell)^2}}, \tag{4.6.7}$$

where $U = \frac{\Gamma}{8\pi\ell}$. Thus, (for $\Gamma > 0$) the vortex starts above the half-plane at $t = -\infty$ at $x_{01} = -\infty$, $x_{02} = 2\ell$ and translates towards the edge, initially at speed U parallel to the plane. It crosses the x_1 axis at $t = 0$ at $x_{01} = \ell$, and proceeds along a symmetrical path below the half-plane.

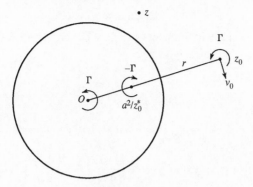

Fig. 4.6.3.

Example 2: Vortex motion outside a cylinder A vortex Γ is located at $z_0 = re^{i\theta}$ outside a rigid cylinder of radius $a(< r)$ with center at the origin (Fig. 4.6.3). There is no net circulation around the cylinder. The complex potential is obtained by placing an image vortex $-\Gamma$ at the *inverse point* $z = a^2/z_0^*$ together with a vortex $+\Gamma$ at the center of the cylinder. The two interior vortices ensure that the total circulation around the cylinder vanishes. Then

$$w(z) = -\frac{i\Gamma}{2\pi} \ln(z - z_0) + \frac{i\Gamma}{2\pi} \ln\left(z - \frac{a^2}{z_0^*}\right) - \frac{i\Gamma}{2\pi} \ln z.$$

The first term on the right is the self-potential of the vortex (in the notation of (4.6.1) $\zeta \equiv z$), so that the equation of motion of the vortex is

$$\frac{dz_0^*}{dt} = \frac{i\Gamma z_0^*}{2\pi(r^2 - a^2)} - \frac{i\Gamma}{2\pi z_0} \equiv \frac{i\Gamma a^2}{2\pi z_0(r^2 - a^2)}.$$

By multiplying by z_0 and adding the complex conjugate equation we see that $r = $ constant, and

$$r\frac{d}{dt}e^{-i\theta} \equiv -ire^{-i\theta}\frac{d\theta}{dt} = \frac{i\Gamma a^2 e^{-i\theta}}{2\pi r(r^2 - a^2)}.$$

Therefore,

$$\frac{d\theta}{dt} = \frac{-\Gamma a^2}{2\pi r^2(r^2 - a^2)},$$

and (for $\Gamma > 0$) the vortex trajectory is a circle traversed in the clockwise direction at speed

$$v_0 = \frac{\Gamma a^2}{2\pi r(r^2 - a^2)}. \tag{4.6.8}$$

Problems 4

1. Show that in inviscid, homentropic flow (where div $\mathbf{v} \neq 0$) the vorticity equation (4.2.7) takes the form

$$\frac{D}{Dt}\left(\frac{\omega}{\rho}\right) = \left(\frac{\omega}{\rho} \cdot \nabla\right)\mathbf{v}.$$

2. Use the relation $\int (y_i \omega_j(\mathbf{y}, t) + y_j \omega_i(\mathbf{y}, t)) \, d^3\mathbf{y} = 0$ to show that

$$\frac{1}{4\pi |\mathbf{x}|^3} \int (\mathbf{x} \cdot \mathbf{y})\boldsymbol{\omega}(\mathbf{y}, t) \, d^3\mathbf{y} = \nabla\left(\frac{1}{4\pi |\mathbf{x}|}\right) \wedge \left[\frac{1}{2}\int \mathbf{y} \wedge \boldsymbol{\omega}(\mathbf{y}, t) \, d^3\mathbf{y}\right].$$

   Deduce the formulae (4.3.3).

3. Calculate the added mass coefficients M_{ij} for an infinite, rigid strip of width $2a$.

4. Calculate the added mass coefficients M_{ij} for an infinite, rigid cylinder of radius a.

5. Calculate the unsteady lift and drag exerted on a rigid circular cylinder of radius a produced by a parallel line vortex of circulation Γ in the presence of a uniform mean flow normal to the cylinder. Assume the motion is ideal and that the net circulation around the cylinder vanishes.

6. Repeat Question 5 under the assumption that the vortex is convected solely by the mean flow (i.e., when the induced component of the motion of Γ produced by image vortices in the cylinder is neglected).

7. A rigid sphere of radius a translates at constant velocity $\mathbf{U} = (U, 0, 0)$, $U > 0$, along the x_1 axis. Use the creeping flow approximation $\omega = \text{curl} \, (3a\mathbf{U}/2|\mathbf{x}|)$, where the coordinate origin is taken at the center of the sphere, to deduce the Stokes drag formula $D = 6\pi \eta U a$.

8. A gas bubble in water is set into translational motion at velocity $\mathbf{U}(t)$ by sound whose wavelength greatly exceeds the bubble radius. If the acoustic particle velocity near the bubble would equal $\mathbf{V}(t)$ in the absence of the bubble, show that $\mathbf{U} = 3\mathbf{V}$ when the mass of the air within the bubble is neglected.

9. Calculate the path of a line vortex of strength Γ that is parallel to a rigid strip occupying $-a < x_1 < a$, $x_2 = 0$, $-\infty < x_3 < \infty$. Assume the fluid is at rest at infinity and that there is no net circulation around the strip. Determine the unsteady force on the strip.

10. Calculate the path of a line vortex of strength Γ that is parallel to a rigid elliptic cylinder of semi-major and minor axes respectively equal to a and b. Assume the fluid is at rest at infinity and that there is no net circulation around the cylinder.

11. A line vortex of strength Γ is adjacent to a rigid right-angle corner whose sides lie along the positive x_1 and x_2 axes, the vortex being parallel to the edge of the corner. Show that the vortex traverses a path with polar representation $r \sin 2\theta = $ constant.

12. Calculate the trajectories of a vortex pair consisting of two parallel line vortices of strengths $\pm\Gamma$ moving under their mutual induction towards a rigid plane parallel to the line of centers of the vortices.

13. Calculate the trajectory of a line vortex of strength Γ adjacent to the rigid half-plane $x_1 < 0$, $x_2 = 0$ in the presence of a uniform mean flow at speed U in the positive x_1 direction.

14. Show that the transformation $\zeta = \sqrt{z^2/a^2 + 1}$, $a > 0$ maps the upper z plane cut by a thin rigid barrier along the imaginary axis between $z = 0$ and $z = ia$ onto the upper ζ plane. Deduce that a line vortex Γ at $z = z_0(t)$ traverses a path determined by the equation

$$\frac{dz_0^*}{dt} = -\frac{i\Gamma}{4\pi}\left\{\frac{a^2}{z_0\left(z_0^2 + a^2\right)} - \frac{2z_0}{\left[z_0^2 + a^2 - |z_0^2 + a^2|\right]}\right\},$$

provided the fluid is at rest at infinity.

z plane

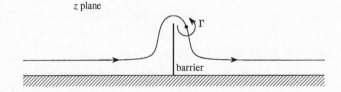

barrier

5

Vortex Sound

5.1 The Role of Vorticity in Lighthill's Theory

At low Mach numbers in unbounded, homentropic flow the value of Lighthill's quadrupole source (2.2.2) can be approximated by means of the Biot–Savart induction formula (4.3.1):

$$T_{ij} \approx \rho_0 u_i u_j, \qquad \mathbf{u}(\mathbf{x}, t) = \mathrm{curl} \int \frac{\boldsymbol{\omega}(\mathbf{y}, t)\, d^3\mathbf{y}}{4\pi\, |\mathbf{x} - \mathbf{y}|}. \qquad (5.1.1)$$

To examine this in more detail, consider the acoustically compact eddy of Fig. 2.2.1 consisting of vorticity of characteristic length ℓ, and take the coordinate origin within the eddy. Put

$$\mathbf{v} = \mathbf{u} + \nabla\varphi;$$

$\mathbf{u}$ involves the whole *incompressible* component of velocity, and $u \sim \mathrm{O}(1/|\mathbf{x}|^3)$ as $|\mathbf{x}| \to \infty$ (see Section 4.3). Because div $\mathbf{u} = 0$, the scalar potential φ describes compressible motions, and the continuity equation becomes

$$\nabla^2\varphi + \frac{1}{\rho}\frac{D\rho}{Dt} = 0.$$

But $p - p_0 \sim \rho_0 u^2$ in the eddy, where the characteristic frequency $\sim u/\ell$. Thus,

$$\frac{Dp}{Dt} \sim \frac{\rho_0 u^3}{\ell}. \quad \text{Hence,} \quad \frac{1}{\rho}\frac{D\rho}{Dt} = \frac{1}{\rho c^2}\frac{Dp}{Dt} \sim \frac{u}{\ell}M^2, \quad M = \frac{u}{c_0}.$$

Hence, in order of magnitude

$$\nabla\varphi = \mathrm{O}(uM^2) \quad \text{within the eddy, where } |\mathbf{x}| \sim \ell. \qquad (5.1.2)$$

114

Now write

$$\frac{\partial^2 (u_i u_j)}{\partial x_i \partial x_j} = \text{div}(\boldsymbol{\omega} \wedge \mathbf{u}) + \nabla^2 \left(\frac{1}{2}u^2\right) \qquad (5.1.3)$$

and express the solution $p(\mathbf{x}, t) = c_0^2(\rho - \rho_0)$ of Lighthill's equation given by (2.2.1) in the form

$$p(\mathbf{x}, t) = p_1(\mathbf{x}, t) + p_2(\mathbf{x}, t),$$

where, using (1.9.6) and (1.9.8) as $|\mathbf{x}| \to \infty$,

$$p_1(\mathbf{x}, t) = \frac{-\rho_0 x_i}{4\pi c_0 |\mathbf{x}|^2} \frac{\partial}{\partial t} \int (\boldsymbol{\omega} \wedge \mathbf{u})_i \left(\mathbf{y}, t - \frac{|\mathbf{x}|}{c_0} + \frac{\mathbf{x} \cdot \mathbf{y}}{c_0 |\mathbf{x}|}\right) d^3\mathbf{y}, \quad (5.1.4)$$

$$p_2(\mathbf{x}, t) = \frac{\rho_0}{4\pi c_0^2 |\mathbf{x}|} \frac{\partial^2}{\partial t^2} \int \frac{1}{2}u^2 \left(\mathbf{y}, t - \frac{|\mathbf{x}|}{c_0} + \frac{\mathbf{x} \cdot \mathbf{y}}{c_0 |\mathbf{x}|}\right) d^3\mathbf{y}. \qquad (5.1.5)$$

When retarded time variations $\mathbf{x} \cdot \mathbf{y}/c_0 |\mathbf{x}|$ within the eddy are neglected the identity (5.1.3) and the divergence theorem imply that $\int \boldsymbol{\omega} \wedge \mathbf{u} \, d^3\mathbf{y} \equiv 0$, because $u \sim O(1/|\mathbf{y}|^3)$ as $|\mathbf{y}| \to \infty$. To estimate the value of the integral in (5.1.4) it is therefore necessary to expand the integrand to the next higher approximation in the retarded time:

$$(\boldsymbol{\omega} \wedge \mathbf{u}) \left(\mathbf{y}, t - \frac{|\mathbf{x}|}{c_0} + \frac{\mathbf{x} \cdot \mathbf{y}}{c_0 |\mathbf{x}|}\right) = (\boldsymbol{\omega} \wedge \mathbf{u}) \left(\mathbf{y}, t - \frac{|\mathbf{x}|}{c_0}\right)$$

$$+ \frac{\mathbf{x} \cdot \mathbf{y}}{c_0 |\mathbf{x}|} \frac{\partial}{\partial t} \left\{(\boldsymbol{\omega} \wedge \mathbf{u}) \left(\mathbf{y}, t - \frac{|\mathbf{x}|}{c_0}\right)\right\} + \cdots.$$

We now find

$$p_1(\mathbf{x}, t) \approx \frac{-\rho_0 x_i x_j}{4\pi c_0^2 |\mathbf{x}|^3} \frac{\partial^2}{\partial t^2} \int y_i (\boldsymbol{\omega} \wedge \mathbf{u})_j \left(\mathbf{y}, t - \frac{|\mathbf{x}|}{c_0}\right) d^3\mathbf{y} \sim \frac{\ell}{|\mathbf{x}|} \rho_0 u^2 M^2,$$

$$|\mathbf{x}| \to \infty. \quad (5.1.6)$$

The order of magnitude of $p_2(\mathbf{x}, t)$ is estimated by using the momentum equation (4.2.3). Because $\text{div} \, \mathbf{v} \sim O(M^2)$ within the source region, we can write

$$\frac{\partial \mathbf{u}}{\partial t} + \nabla \left(\int \frac{dp}{\rho} + \frac{1}{2}v^2 + \frac{\partial \varphi}{\partial t}\right) = -\boldsymbol{\omega} \wedge \mathbf{u} - \boldsymbol{\omega} \wedge \nabla\varphi - \nu \, \text{curl} \, \boldsymbol{\omega}.$$

Take the scalar product with $\mathbf{u}$

$$\frac{\partial}{\partial t}\left(\frac{1}{2}u^2\right) + \text{div}\left\{\mathbf{u}\left(\int \frac{dp}{\rho} + \frac{1}{2}v^2 + \frac{\partial \varphi}{\partial t}\right)\right\}$$

$$= -\mathbf{u} \cdot \boldsymbol{\omega} \wedge \nabla\varphi - \nu \mathbf{u} \cdot \text{curl} \, \boldsymbol{\omega}$$

$$= -\mathbf{u} \cdot \boldsymbol{\omega} \wedge \nabla\varphi + \nu(\text{div}\,(\mathbf{u} \wedge \boldsymbol{\omega}) - \omega^2)$$

and integrate over the whole of space. The contributions from the divergence terms vanish because $\mathbf{u}(\int dp/\rho + \frac{1}{2}v^2 + \partial\varphi/\partial t)$ tends to zero at least as fast as $1/|\mathbf{y}|^3$ as $|\mathbf{y}| \to \infty$, where also $\boldsymbol{\omega} = 0$. Hence, using the estimate (5.1.2)

$$\frac{\partial}{\partial t} \int \frac{1}{2} u^2(\mathbf{y}, t)\, d^3\mathbf{y} = -\int (\mathbf{u} \cdot \boldsymbol{\omega} \wedge \nabla\varphi + v\omega^2)(\mathbf{y}, t)\, d^3\mathbf{y} \sim \ell^2 u^3 M^2 + \frac{\ell^2 u^3}{\mathrm{Re}},$$

(5.1.7)

where $\mathrm{Re} = u\ell/v$ typically exceeds 10^4 in turbulent flow. The two terms on the right-hand side nominally represent the dissipation of the turbulent motions respectively by acoustic radiation and by viscous damping. (We have already seen, however, in Chapter 2, Equation (2.2.5) that a more accurate estimate of the radiation damping is $\ell^2 u^3 M^5$.)

Thus, when retarded time variations are neglected in (5.1.5), we find

$$p_2(\mathbf{x}, t) \sim \frac{\ell}{|\mathbf{x}|} \rho_0 u^2 M^4 + \frac{\ell}{|\mathbf{x}|} \frac{\rho_0 u^2 M^2}{\mathrm{Re}},$$

and therefore that $p_2 \ll p_1$ in turbulent flow where $M \ll 1$ and $\mathrm{Re} \gg 1$.

We conclude that the component

$$\mathrm{div}(\rho_0 \boldsymbol{\omega} \wedge \mathbf{v}) \quad \text{of the Lighthill quadrupole } \frac{\partial^2(\rho_0 v_i v_j)}{\partial x_i \partial x_j}$$

is principal source of sound at low Mach numbers.

5.2 The Equation of Vortex Sound

Lighthill's equation (2.1.12) can be recast in a form that emphasizes the prominent rôle of vorticity in the production of sound by taking the **total enthalpy**

$$B = \int \frac{dp}{\rho} + \frac{1}{2} v^2$$

as the independent acoustic variable, in place of Lighthill's $c_0^2(\rho - \rho_0)$. The total enthalpy occurs naturally in Crocco's form (4.2.3) of the momentum equation. In the following we shall actually use the Approximation (4.2.5) of this equation, in which the viscous term $\frac{4}{3} v \nabla(\mathrm{div}\, \mathbf{v})$ is neglected. Indeed, the principal effect of this term is to attenuate the sound once it has been generated and is propagating to a distant observer in the source-free region of the flow. This attenuation can be significant in applications, but is of no particular interest when studying sound generation mechanisms. All viscous stresses can be ignored in a high Reynolds number source flow *except* possibly within surface boundary layers on bodies immersed in the flow. But surface friction is dominated by the vorticity term $-\nu \operatorname{curl} \boldsymbol{\omega}$, which is retained in (4.2.5).

In irrotational flow Crocco's equation (4.2.5) reduces to

$$\frac{\partial \mathbf{v}}{\partial t} = -\nabla B.$$

In other words,

$$B = -\frac{\partial \varphi}{\partial t} \quad \text{in regions where } \boldsymbol{\omega} = 0, \tag{5.2.1}$$

where $\varphi(\mathbf{x}, t)$ is the velocity potential that determines the whole motion in the irrotational regions of the fluid. B is therefore *constant* in steady irrotational flow, and at large distances from the acoustic sources perturbations in B represent acoustic waves.

If the mean flow is at rest in the far field, the acoustic pressure is given by

$$p = \rho_0 B \equiv -\rho_0 \frac{\partial \varphi}{\partial t}. \tag{5.2.2}$$

To calculate the pressure in terms of B elsewhere in the flow, we use the definition

$$\int \frac{dp}{\rho} = B - \frac{1}{2} v^2.$$

Differentiating with respect to time and using Crocco's equation (4.2.5), we have

$$
\begin{aligned}
\frac{1}{\rho} \frac{\partial p}{\partial t} &= \frac{\partial B}{\partial t} - \mathbf{v} \cdot \frac{\partial \mathbf{v}}{\partial t} \\
&= \frac{\partial B}{\partial t} - \mathbf{v} \cdot (-\nabla B - \boldsymbol{\omega} \wedge \mathbf{v} - \nu \operatorname{curl} \boldsymbol{\omega}) \\
&= \frac{DB}{Dt} + \nu \mathbf{v} \cdot \operatorname{curl} \boldsymbol{\omega}.
\end{aligned}
$$

The small viscous correction can be ignored in high Reynolds number source flows, where p and B can be taken to be related by

$$\frac{1}{\rho} \frac{\partial p}{\partial t} = \frac{DB}{Dt}. \tag{5.2.3}$$

5.2.1 Reformulation of Lighthill's Equation

Multiply Crocco's equation (4.2.5) by the density ρ and take the divergence

$$\operatorname{div}\left(\rho \frac{\partial \mathbf{v}}{\partial t}\right) + \nabla \cdot (\rho \nabla B) = -\operatorname{div}(\rho \boldsymbol{\omega} \wedge \mathbf{v}). \tag{5.2.4}$$

The first term on the left is expressed in terms of B by using the continuity equation in the form

$$\text{div } \mathbf{v} = -\frac{1}{\rho}\frac{D\rho}{Dt},$$

and writing

$$\text{div}\left(\rho\frac{\partial \mathbf{v}}{\partial t}\right) = \nabla\rho \cdot \frac{\partial \mathbf{v}}{\partial t} + \rho\frac{\partial}{\partial t}\text{div }\mathbf{v}$$

$$= \nabla\rho \cdot \frac{\partial \mathbf{v}}{\partial t} - \rho\frac{\partial}{\partial t}\left(\frac{1}{\rho}\frac{D\rho}{Dt}\right)$$

$$= \nabla\rho \cdot \frac{\partial \mathbf{v}}{\partial t} - \rho\frac{\partial}{\partial t}\left(\frac{1}{\rho}\frac{\partial\rho}{\partial t}\right) - \frac{\partial \mathbf{v}}{\partial t}\cdot\nabla\rho - \rho\mathbf{v}\cdot\nabla\left(\frac{1}{\rho}\frac{\partial\rho}{\partial t}\right)$$

$$= -\rho\frac{D}{Dt}\left(\frac{1}{\rho}\frac{\partial\rho}{\partial t}\right)$$

$$= -\rho\frac{D}{Dt}\left(\frac{1}{\rho c^2}\frac{\partial p}{\partial t}\right)$$

$$= -\rho\frac{D}{Dt}\left(\frac{1}{c^2}\frac{DB}{Dt}\right),$$

where Equation (5.2.3) has been used on the last line. Substituting into (5.2.4) and dividing by ρ, we obtain the desired **vortex sound equation** for homentropic flow

$$\left(\frac{D}{Dt}\left(\frac{1}{c^2}\frac{D}{Dt}\right) - \frac{1}{\rho}\nabla\cdot(\rho\nabla)\right)B = \frac{1}{\rho}\text{div}(\rho\boldsymbol{\omega}\wedge\mathbf{v}). \qquad (5.2.5)$$

The vortex source on the right-hand side vanishes in irrotational regions; if $\boldsymbol{\omega} = 0$ everywhere, and if there are no moving boundaries, the total enthalpy B is constant, and there are no sound waves propagating in the fluid. If acoustic waves cannot enter from infinity, it follows that the (homentropic) flow can generate sound only if moving vorticity is present, and the right-hand side of (5.2.5) may be identified as the analytical representation of the acoustic sources. The differential operator on the left describes propagation of the sound through the nonuniform flow; as in the case of Lighthill's equation, when the source region is very extensive it will not normally be permissible to neglect the interaction of the sound with the vorticity through which it propagates.

5.2.2 Sound Waves in Irrotational Mean Flow

Let an irrotational mean flow be defined by the velocity potential $\varphi_0(\mathbf{x})$, with mean velocity $\mathbf{U} = \nabla\varphi_0$. In an unbounded fluid $\mathbf{U} = $ constant; the mean velocity can vary with position only if the fluid is bounded, either internally by an airfoil, say, or externally by the walls of a duct of variable cross section ($\varphi_0(\mathbf{x})$ can be multiple-valued if the boundaries are multiply connected, but the mean velocity is always single-valued).

Consider an irrotational disturbance $\varphi'(\mathbf{x}, t)$, and set

$$\varphi(\mathbf{x}, t) = \varphi_0(\mathbf{x}) + \varphi'(\mathbf{x}, t).$$

It can be shown that the general, nonlinear equation satisfied by φ is

$$\frac{1}{c^2}\frac{\partial^2 \varphi}{\partial t^2} + \frac{1}{c^2}\frac{D}{Dt}\left(\frac{1}{2}(\nabla\varphi)^2\right) + \frac{1}{c^2}\frac{\partial}{\partial t}\left(\frac{1}{2}(\nabla\varphi)^2\right) - \nabla^2\varphi = 0.$$

The linearized version of this equation describes the propagation of small amplitude sound waves determined by $\varphi'(\mathbf{x}, t)$. However, when $\omega = 0$ the linearized equation for $B = -\partial\varphi'/\partial t \equiv -\dot{\varphi}$ is more easily derived from (5.2.5), which becomes

$$\left(\frac{D}{Dt}\left(\frac{1}{c^2}\frac{D}{Dt}\right) - \frac{1}{\rho}\nabla\cdot(\rho\nabla)\right)\dot{\varphi} = 0. \tag{5.2.6}$$

The coefficients of the differential operators in this equation are functions of both mean and perturbation quantities, but the linearized equation is obtained merely by replacing these coefficients by their values in the *absence of the sound*. In homentropic flow the mean density and sound speed can be expressed in terms of the variable mean velocity $\mathbf{U}(\mathbf{x})$, and

$$\frac{D}{Dt} \approx \frac{\partial}{\partial t} + \mathbf{U}\cdot\nabla.$$

Furthermore, because the mean flow does not depend on time, we can take the perturbation potential φ', rather than $\dot{\varphi}$, as the acoustic variable. The linearized equation then becomes

$$\left\{\left(\frac{\partial}{\partial t} + \mathbf{U}\cdot\nabla\right)\left[\frac{1}{c^2}\left(\frac{\partial}{\partial t} + \mathbf{U}\cdot\nabla\right)\right] - \frac{1}{\rho}\nabla\cdot(\rho\nabla)\right\}\varphi' = 0,$$

where $c \equiv c(\mathbf{x})$ and $\rho \equiv \rho(\mathbf{x})$ are the local sound speed and density in the steady flow.

5.2.3 Vortex Sound at Low Mach Numbers

When the characteristic Mach number M is small the local mean values of the density and sound speed are related to their uniform respective values ρ_0 and

c_0 at infinity by relations of the form

$$\frac{c}{c_0} \sim 1 + O(M^2), \qquad \frac{\rho}{\rho_0} \sim 1 + O(M^2).$$

The vortex sound equation (5.2.5) can therefore be simplified by (a) taking $c = c_0$, and $\rho = \rho_0$, and (b) by neglecting nonlinear effects of propagation and the scattering of sound by the vorticity. The production of sound is then governed by the simpler equation

$$\left(\frac{1}{c_0^2} \frac{\partial^2}{\partial t^2} - \nabla^2 \right) B = \text{div}(\boldsymbol{\omega} \wedge \mathbf{v}), \qquad (5.2.7)$$

and in the far field the acoustic pressure is given by the linearized approximation

$$p(\mathbf{x}, t) \approx \rho_0 B(\mathbf{x}, t). \qquad (5.2.8)$$

5.2.4 Example 1 (Powell 1963): Sound Generation by a Spinning Vortex Pair

Two parallel vortex filaments each of circulation Γ and distance 2ℓ apart rotate about the x_3 axis midway between them (Fig. 5.2.1) at angular velocity $\Omega = \Gamma/4\pi\ell^2$, provided the Mach number is small enough for the motion to be regarded as incompressible. Their positions at time t are

$$\bar{\mathbf{x}} = (x_1, x_2) = \pm \mathbf{s} \equiv \pm(s_1(t), s_2(t)) = \pm\ell(\cos \Omega t, \sin \Omega t).$$

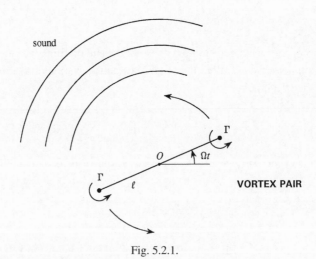

Fig. 5.2.1.

The vorticity distribution is

$$\omega = \Gamma \mathbf{k}(\delta(\bar{\mathbf{x}} - \mathbf{s}) + \delta(\bar{\mathbf{x}} + \mathbf{s})),$$

where $\mathbf{k}$ is a unit vector in the x_3 direction, parallel to the vortices. The vortex convection velocities are

$$\mathbf{v} = \pm \Omega \mathbf{k} \wedge \mathbf{s}(t) \quad \text{at} \quad \bar{\mathbf{x}} = \pm \mathbf{s}(t), \quad \text{where } U = \Omega \ell \ll c_0.$$

Hence, because $\mathbf{k} \wedge (\mathbf{k} \wedge \mathbf{s}) = -\mathbf{s}$,

$$\omega \wedge \mathbf{v} = -\Gamma \Omega \mathbf{s}(t)[\delta(\bar{\mathbf{x}} - \mathbf{s}) - \delta(\bar{\mathbf{x}} + \mathbf{s})].$$

If this is expanded in powers of the radius ℓ of the circular orbit (and it can be verified that this is equivalent to expanding the acoustic pressure in powers of $M = U/c_0 \ll 1$) the ith component of the first nonzero term is

$$(\omega \wedge \mathbf{v})_i \approx \frac{\partial}{\partial \bar{x}_j}(2\Gamma \Omega s_i(t)s_j(t)\delta(\bar{\mathbf{x}})),$$

so that the vortex sound source is equivalent to the quadrupole

$$\operatorname{div}(\omega \wedge \mathbf{v}) \approx \frac{\partial^2}{\partial \bar{x}_i \partial \bar{x}_j}(2\Gamma \Omega s_i(t)s_j(t)\delta(\bar{\mathbf{x}})).$$

The solution of the vortex sound equation (5.2.7) for this quadrupole source is (c.f., (2.2.1))

$$B = \frac{1}{4\pi} \frac{\partial^2}{\partial \bar{x}_i \partial \bar{x}_j} \int 2\Gamma \Omega \, (s_i s_j) \left(t - \frac{|\mathbf{x} - \mathbf{y}|}{c_0} \right) \frac{\delta(\bar{\mathbf{y}}) \, d^3 \mathbf{y}}{|\mathbf{x} - \mathbf{y}|}, \quad \bar{\mathbf{y}} = (y_1, y_2).$$

In the acoustic far field, we use (1.9.7):

$$\frac{\partial}{\partial \bar{x}_j} \approx \frac{-\bar{x}_j}{c_0(r^2 + (x_3 - y_3)^2)^{\frac{1}{2}}} \frac{\partial}{\partial t},$$

where $r = (x_1^2 + x_2^2)^{\frac{1}{2}}$ is the perpendicular distance from the centroid of the vortices (the x_3 axis). Thus, by setting $\xi = y_3 - x_3$, we can write in the acoustic far field, where $B \approx p/\rho_0$,

$$p \approx \frac{\rho_0 \Gamma \Omega \bar{x}_i \bar{x}_j}{2\pi c_0^2} \frac{\partial^2}{\partial t^2} \int_{-\infty}^{\infty} (s_i s_j) \left(t - \frac{(r^2 + \xi^2)^{\frac{1}{2}}}{c_0} \right) \frac{d\xi}{(r^2 + \xi^2)^{\frac{3}{2}}}, \quad r \to \infty.$$

$$(5.2.9)$$

In this formula

$$(s_i s_j)(t) = \frac{\ell^2}{2} \begin{pmatrix} 1 + \cos 2\Omega t & \sin 2\Omega t \\ \sin 2\Omega t & 1 - \cos 2\Omega t \end{pmatrix},$$

but the constant terms in the matrix can be omitted because of the time derivatives in (5.2.9). The integration in (5.2.9) can now be performed by the approximate method described below in Example 2 in the limit that $\Omega r/c_0 \to \infty$, i.e., in the limit in which the radial distance r greatly exceeds the acoustic wavelength

$$\int_{-\infty}^{\infty} (s_i s_j) \left(t - \frac{(r^2 + \xi^2)^{\frac{1}{2}}}{c_0} \right) \frac{d\xi}{(r^2 + \xi^2)^{\frac{3}{2}}}$$

$$\approx \frac{\ell^2}{2r^2} \left(\frac{\pi c_0}{\Omega r} \right)^{\frac{1}{2}} \begin{pmatrix} \cos\left(2\Omega[t] - \frac{\pi}{4}\right) & \sin\left(2\Omega[t] - \frac{\pi}{4}\right) \\ \sin\left(2\Omega[t] - \frac{\pi}{4}\right) & -\cos\left(2\Omega[t] - \frac{\pi}{4}\right) \end{pmatrix},$$

where $[t] = t - r/c_0$. Hence, introducing polar coordinates $\bar{\mathbf{x}} = r(\cos\theta, \sin\theta)$ we find

$$\bar{x}_i \bar{x}_j \frac{\ell^2}{2r^2} \left(\frac{\pi c_0}{\Omega r} \right)^{\frac{1}{2}} \begin{pmatrix} \cos\left(2\Omega[t] - \frac{\pi}{4}\right) & \sin\left(2\Omega[t] - \frac{\pi}{4}\right) \\ \sin\left(2\Omega[t] - \frac{\pi}{4}\right) & -\cos\left(2\Omega[t] - \frac{\pi}{4}\right) \end{pmatrix}_{ij}$$

$$= \frac{\ell^2}{2} \left(\frac{\pi c_0}{\Omega r} \right)^{\frac{1}{2}} (\cos\theta, \sin\theta) \begin{pmatrix} \cos\left(2\Omega[t] - \frac{\pi}{4}\right) & \sin\left(2\Omega[t] - \frac{\pi}{4}\right) \\ \sin\left(2\Omega[t] - \frac{\pi}{4}\right) & -\cos\left(2\Omega[t] - \frac{\pi}{4}\right) \end{pmatrix} \begin{pmatrix} \cos\theta \\ \sin\theta \end{pmatrix}$$

$$= \frac{\ell^2}{2} \left(\frac{\pi c_0}{\Omega r} \right)^{\frac{1}{2}} \cos\left(2\theta - 2\Omega[t] + \frac{\pi}{4}\right),$$

and, therefore, (5.2.9) becomes

$$p \approx \frac{-\rho_0 \Gamma \Omega^3 \ell^2}{\pi c_0^2} \left(\frac{\pi c_0}{\Omega r} \right)^{\frac{1}{2}} \cos\left[2\theta - 2\Omega \left(t - \frac{r}{c_0} \right) + \frac{\pi}{4} \right]$$

$$= -4\sqrt{\frac{\pi \ell}{r}} \rho_0 U^2 M^{3/2} \cos\left[2\theta - 2\Omega \left(t - \frac{r}{c_0} \right) + \frac{\pi}{4} \right], \quad \frac{\Omega r}{c_0} \to \infty.$$

$$(5.2.10)$$

The amplitude of the sound decreases like $1/\sqrt{r}$ (instead of $1/r$) because the waves are spreading cylindrically in two dimensions. The sound power must now be calculated by considering the integral

$$\int_{\Sigma} \frac{p^2}{\rho_0 c_0} \, dS$$

over the surface of a large circular cylinder $r = $ constant. Taking the time average, we can show that the acoustic power *per unit length* of the vortices $\sim \ell \rho_0 U^3 M^4$. This differs by a factor of the Mach number M from the power radiated by a compact body of three-dimensional turbulence, and is characteristic of the acoustic power produced by two-dimensional regions of turbulence *in an unbounded fluid*.

5.2.5 Example 2

Show that

$$\int_{-\infty}^{\infty} f\left(\frac{\xi}{r}\right) e^{i\kappa_0 \sqrt{r^2+\xi^2}} \, d\xi \approx r f(0) \left(\frac{2\pi}{\kappa_0 r}\right)^{\frac{1}{2}} e^{i\left(\kappa_0 r + \frac{\pi}{4}\right)}, \quad \kappa_0 r \to \infty. \quad (5.2.11)$$

Put $\xi = \mu r$, then

$$I \equiv \int_{-\infty}^{\infty} f\left(\frac{\xi}{r}\right) e^{i\kappa_0 \sqrt{r^2+\xi^2}} \, d\xi = r \int_{-\infty}^{\infty} f(\mu) e^{i\kappa_0 r \sqrt{1+\mu^2}} \, d\mu.$$

As $\kappa_0 r \to \infty$ the exponential factor oscillates increasingly rapidly, and the main contribution to the integral is from the neighborhood of that value of μ where the oscillations are stationary. This occurs at $\mu = 0$. The integrand is therefore expanded about this point. In the first approximation, $f(\mu)$ can be replaced by $f(0)$, and

$$e^{i\kappa_0 r \sqrt{1+\mu^2}} \approx e^{i\kappa_0 r + i\kappa_0 r \mu^2/2}.$$

Thus,

$$I = r \int_{-\infty}^{\infty} f(\mu) e^{i\kappa_0 r \sqrt{1+\mu^2}} \, d\mu \approx r f(0) e^{i\kappa_0 r} \int_{-\infty}^{\infty} e^{i\kappa_0 r \mu^2/2} \, d\mu$$

$$= r f(0) e^{i\kappa_0 r} \left(\frac{2\pi}{\kappa_0 r}\right)^{\frac{1}{2}} e^{\frac{i\pi}{4}},$$

which yields (5.2.11).

In particular,

$$\int_{-\infty}^{\infty} \left\{ \cos 2\Omega \left(t - \frac{(r^2+\xi^2)^{\frac{1}{2}}}{c_0} \right) - i \sin 2\Omega \left(t - \frac{(r^2+\xi^2)^{\frac{1}{2}}}{c_0} \right) \right\} \frac{d\xi}{(r^2+\xi^2)^{\frac{3}{2}}}$$

$$= e^{-2i\Omega t} \int_{-\infty}^{\infty} e^{i\frac{2\Omega}{c_0}\sqrt{r^2+\xi^2}} \frac{d\xi}{(r^2+\xi^2)^{\frac{3}{2}}}$$

$$\approx \frac{1}{r^2} \left(\frac{\pi c_0}{\Omega r}\right)^{\frac{1}{2}} e^{-i[2\Omega(t-\frac{r}{c_0}) - \frac{\pi}{4}]}, \quad \frac{\Omega r}{c_0} \to \infty.$$

Hence,

$$
\int_{-\infty}^{\infty} \cos 2\Omega \left(t - \frac{(r^2 + \xi^2)^{\frac{1}{2}}}{c_0} \right) \frac{d\xi}{(r^2 + \xi^2)^{\frac{3}{2}}}
$$

$$
\approx \frac{1}{r^2} \left(\frac{\pi c_0}{\Omega r} \right)^{\frac{1}{2}} \cos \left[2\Omega \left(t - \frac{r}{c_0} \right) - \frac{\pi}{4} \right]
$$

$$
\int_{-\infty}^{\infty} \sin 2\Omega \left(t - \frac{(r^2 + \xi^2)^{\frac{1}{2}}}{c_0} \right) \frac{d\xi}{(r^2 + \xi^2)^{\frac{3}{2}}}
$$

$$
\approx \frac{1}{r^2} \left(\frac{\pi c_0}{\Omega r} \right)^{\frac{1}{2}} \sin \left[2\Omega \left(t - \frac{r}{c_0} \right) - \frac{\pi}{4} \right].
$$

5.3 Vortex–Surface Interaction Noise

The small Mach number vortex sound equation (5.2.7) is now applied to determine the sound generated by vorticity in the neighborhood of a fixed body whose surface S may be vibrating at small amplitude (Fig. 5.3.1). The development here is analogous to the derivation in Section 2.3 of Curle's equation.

Introduce a *stationary*, closed control surface S_+ on which $f(\mathbf{x}) = 0$, such that $f(\mathbf{x}) \gtrless 0$ according as $\mathbf{x}$ lies without or within S_+. The body is assumed to be within S_+, and S_+ will subsequently be allowed to shrink down to coincide with the body surface S. Multiply equation (5.2.7) by $H \equiv H(f)$ and

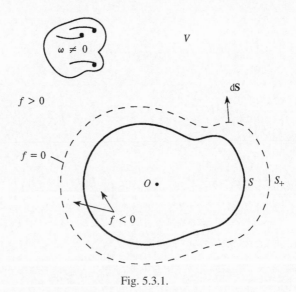

Fig. 5.3.1.

form the inhomogeneous wave equation for the new variable HB. We use the transformations

$$H\nabla^2 B \equiv H \operatorname{div}(\nabla B) = \operatorname{div}(H\nabla B) - \nabla H \cdot \nabla B$$
$$= \nabla^2(HB) - \operatorname{div}(B\nabla H) - \nabla H \cdot \nabla B, \quad (5.3.1)$$
$$\text{and } H \operatorname{div}(\omega \wedge \mathbf{v}) = \operatorname{div}(H\omega \wedge \mathbf{v}) - \nabla H \cdot \omega \wedge \mathbf{v}. \quad (5.3.2)$$

Then, (5.2.7) becomes

$$\left(\frac{1}{c_0^2}\frac{\partial^2}{\partial t^2} - \nabla^2\right)(HB) = -\operatorname{div}(B \nabla H) - \nabla H \cdot (\nabla B + \omega \wedge \mathbf{v}) + \operatorname{div}(H\omega \wedge \mathbf{v}).$$

This equation is formally valid everywhere, including the region within S_+ where $HB \equiv 0$. The source terms involving ∇H are concentrated on the control surface. When $\mathbf{x}$ lies in the exterior region these surface terms take account of the presence of the solid body inside S_+; if the body is absent (so that S_+ is filled with fluid), the surface sources constitute a representation 'to the outside world' in $f > 0$ of the various hydrodynamic or acoustic processes that may be occurring within S_+.

Using Crocco's equation (4.2.5), we can make the substitution

$$\nabla H \cdot (\nabla B + \omega \wedge \mathbf{v}) = -\nabla H \cdot \left(\frac{\partial \mathbf{v}}{\partial t} + v\operatorname{curl}\omega\right)$$

$$\equiv -\nabla H \cdot \frac{\partial \mathbf{v}}{\partial t} + v\operatorname{div}(\nabla H \wedge \omega).$$

Hence the vortex sound equation becomes

$$\left(\frac{1}{c_0^2}\frac{\partial^2}{\partial t^2} - \nabla^2\right)(HB) = -\operatorname{div}(B\nabla H) + \nabla H \cdot \frac{\partial \mathbf{v}}{\partial t}$$
$$+ \operatorname{div}(H\omega \wedge \mathbf{v}) - v\operatorname{div}(\nabla H \wedge \omega). \quad (5.3.3)$$

The sources on the right of this equation are either concentrated on the control surface S_+ or lie in the fluid outside S_+; they completely determine B outside this control surface. The solution in this region can therefore be found by using any Green's function $G(\mathbf{x}, \mathbf{y}, t - \tau)$ that satisfies

$$\left(\frac{1}{c_0^2}\frac{\partial^2}{\partial t^2} - \nabla^2\right)G = \delta(\mathbf{x} - \mathbf{y})\delta(t - \tau), \quad \text{where } G = 0 \quad \text{for } t < \tau$$

for $\mathbf{x}$ and $\mathbf{y}$ anywhere within the fluid. In Fig. 5.3.1 the fluid occupies the region V *outside* the surface S of the solid body; the control surface S_+ ($f(\mathbf{x}) = 0$) therefore lies *within* V.

Thus, for points $\mathbf{x}$ within the fluid the solution of (5.3.3) is

$$HB(\mathbf{x}, t) = \int_{-\infty}^{\infty} \int_V G(\mathbf{x}, \mathbf{y}, t - \tau) \left\{ -\text{div}(B\nabla H) + \nabla H \cdot \frac{\partial \mathbf{v}}{\partial \tau} + \text{div}(H\boldsymbol{\omega} \wedge \mathbf{v}) \right.$$

$$\left. - \nu \,\text{div}(\nabla H \wedge \boldsymbol{\omega}) \right\} d^3\mathbf{y} \, d\tau,$$

where all of the source terms within the brace brackets are functions of $\mathbf{y}$ and τ. Those involving ∇H vanish except on the control surface S_+. The divergence terms are removed by application of the divergence theorem, and then using the formula

$$\int_V (\cdot) \nabla H \, d^3\mathbf{y} = \oint_{S_+} (\cdot) \, d\mathbf{S}$$

(see Section 2.3). Let us illustrate the procedure for the first term in the brace brackets of the integrand

$$\int_V G\{-\text{div}(B\nabla H)\} \, d^3\mathbf{y} = - \int_V \{\text{div}(GB\nabla H) - B\nabla G \cdot \nabla H\} \, d^3\mathbf{y}$$

$$= \oint_{S+\Sigma} GB\nabla H \cdot d\mathbf{S} + \int_V B\nabla G \cdot \nabla H \, d^3\mathbf{y}$$

$$= 0 + \oint_{S_+} B\nabla G \cdot d\mathbf{S}$$

$$\equiv \oint_{S_+} B(\mathbf{y}, \tau) \frac{\partial G}{\partial y_n}(\mathbf{x}, \mathbf{y}, t - \tau) \, dS(\mathbf{y}),$$

where all vector operators are with respect to the $\mathbf{y}$ dependence, and Σ is a large, closed 'surface at infinity' where $\boldsymbol{\omega} = 0$. There are no contributions from S and Σ because $\nabla H = 0$ everywhere except on S_+.

The general solution in the region $f > 0$ outside S_+ accordingly becomes

$$B(\mathbf{x}, t) = \oint_{S_+} \left(B(\mathbf{y}, \tau) \frac{\partial G}{\partial y_n}(\mathbf{x}, \mathbf{y}, t - \tau) + G(\mathbf{x}, \mathbf{y}, t - \tau) \frac{\partial v_n}{\partial \tau}(\mathbf{y}, \tau) \right) dS(\mathbf{y}) \, d\tau$$

$$- \int_V H(f(\mathbf{y}))(\boldsymbol{\omega} \wedge \mathbf{v})(\mathbf{y}, \tau) \cdot \frac{\partial G}{\partial \mathbf{y}}(\mathbf{x}, \mathbf{y}, t - \tau) \, d^3\mathbf{y} \, d\tau$$

$$+ \nu \oint_{S_+} \boldsymbol{\omega}(\mathbf{y}, \tau) \wedge \frac{\partial G}{\partial \mathbf{y}}(\mathbf{x}, \mathbf{y}, t - \tau) \cdot dS(\mathbf{y}) \, d\tau, \qquad (5.3.4)$$

where for brevity we have omitted the integration sign for τ, which is understood to vary over the range $-\infty < \tau < \infty$.

We now choose G to have vanishing normal derivative on the surface S of the body. When this is done the control surface S_+ is allowed to shrink down onto S (whereupon the first term in the first integral of (5.3.4) vanishes), and the general solution in the fluid becomes

$$B(\mathbf{x}, t) = - \int_V (\boldsymbol{\omega} \wedge \mathbf{v})(\mathbf{y}, \tau) \cdot \frac{\partial G}{\partial \mathbf{y}}(\mathbf{x}, \mathbf{y}, t - \tau) \, d^3\mathbf{y} \, d\tau + \nu \oint_S \boldsymbol{\omega}(\mathbf{y}, \tau)$$

$$\wedge \frac{\partial G}{\partial \mathbf{y}}(\mathbf{x}, \mathbf{y}, t - \tau) \cdot d\mathbf{S}(\mathbf{y}) \, d\tau + \oint_S G(\mathbf{x}, \mathbf{y}, t - \tau) \frac{\partial v_n}{\partial \tau}(\mathbf{y}, \tau) \, d\mathbf{S}(\mathbf{y}) \, d\tau,$$

$$\text{where} \quad \frac{\partial G}{\partial y_n}(\mathbf{x}, \mathbf{y}, t - \tau) = 0 \quad \text{on } S. \quad (5.3.5)$$

In the acoustic far field ($|\mathbf{x}| \to \infty$) we can replace $B(\mathbf{x}, t)$ by $p(\mathbf{x}, t)/\rho_0$.

The first integral represents the production of sound by vortex sources distributed within the fluid. Green's function takes full account of the influence of the body on the efficiency with which these sources generate sound. The second, surface integral involving the surface vorticity is the contribution from frictional forces on S. To interpret the final term, recall that the control surface S_+ was taken to be fixed in space. This means that fluid can flow through the surface. When S_+ shrinks down to S the implication is that S is also fixed in space. However, the normal velocity v_n can still be nonzero if the surface of the body is *vibrating* at small amplitude, and this term in the solution is actually identical with that given previously in (3.8.2) for a vibrating body in the absence of vortex sources.

In connection with this, it should also be noted that the reciprocal theorem implies that the normal derivative conditions

$$\frac{\partial G}{\partial y_n}(\mathbf{x}, \mathbf{y}, t - \tau) = 0, \quad \frac{\partial G}{\partial x_n}(\mathbf{x}, \mathbf{y}, t - \tau) = 0 \quad \text{respectively for } \mathbf{y}, \mathbf{x} \text{ on } S,$$

are always satisfied *simultaneously*.

The contribution to the sound from surface friction (the first surface integral on the right of (5.3.5)) is nominally of order

$$\frac{1}{\mathrm{Re}} \ll 1, \quad \mathrm{Re} = \frac{v\ell}{\nu},$$

relative to the contribution from the volume vorticity (the first integral), where ℓ is the characteristic length scale of the turbulence or body and v is a typical velocity. At high Reynolds numbers the surface term can therefore be discarded, and in the important case in which the body does not vibrate the acoustic far field is then given by

$$\frac{p}{\rho_0}(\mathbf{x}, t) = -\int_V (\boldsymbol{\omega} \wedge \mathbf{v})(\mathbf{y}, \tau) \cdot \frac{\partial G}{\partial \mathbf{y}}(\mathbf{x}, \mathbf{y}, t - \tau)\, d^3\mathbf{y}\, d\tau. \qquad (5.3.6)$$

5.4 Radiation from an Acoustically Compact Body

When the surface S is acoustically compact the compact Green's function (3.9.1) can be used to evaluate the general solution (5.3.5) *in the far field*. When $|\mathbf{x}| \to \infty$ and the origin is within or close to S, we proceed as already described in Section 3.8 by expanding Green's function to first order in the retarded time across S:

$$G(\mathbf{x}, \mathbf{y}, t - \tau) = \frac{1}{4\pi |\mathbf{X} - \mathbf{Y}|} \delta\left(t - \tau - \frac{|\mathbf{X} - \mathbf{Y}|}{c_0}\right)$$

$$\approx \frac{1}{4\pi |\mathbf{x}|} \delta\left(t - \tau - \frac{|\mathbf{x}|}{c_0}\right) + \frac{x_j Y_j}{4\pi c_0 |\mathbf{x}|^2} \delta'\left(t - \tau - \frac{|\mathbf{x}|}{c_0}\right),$$

$$|\mathbf{x}| \to \infty,$$

where the prime denotes differentiation with respect to t. The first term in this approximation, which is independent of $\mathbf{y}$, clearly makes no contribution to the first two integrals in (5.3.5). It makes a contribution to the final surface integral only if the volume of the body is pulsating. When this happens the resulting monopole radiation from the body is usually large compared to all other sources. We shall, therefore, ignore this possibility, and consider only surface vibrations for which the volume of S is constant; in particular we shall assume that S vibrates as a rigid body. In this case, therefore, the first approximation in the Green's function expansion can again be discarded.

Substituting into (5.3.5) and performing the integrations with respect to τ, we find in the acoustic far field (where $B = p/\rho_0$)

$$p(\mathbf{x}, t) \approx \frac{-\rho_0 x_j}{4\pi c_0 |\mathbf{x}|^2} \frac{\partial}{\partial t} \left[\int (\boldsymbol{\omega} \wedge \mathbf{v}) \cdot \nabla Y_j \, d^3\mathbf{y} - \nu \oint_S \boldsymbol{\omega} \wedge \nabla Y_j \cdot d\mathbf{S}(\mathbf{y})\right.$$

$$\left. - \oint_S \frac{\partial U_n}{\partial t} Y_j \, dS(\mathbf{y})\right], \qquad (5.4.1)$$

where the large square brackets ([ ]) denote that the enclosed quantity is to be evaluated at the retarded time $t - |\mathbf{x}|/c_0$, and U_n is the normal component of velocity of vibration of S.

When the body executes translational oscillations at velocity $\mathbf{U}(t)$ we have

$$U_n = U_i n_i,$$

where n_i is the ith component of the surface normal directed into the fluid. Then

$$\rho_0 \oint_S \frac{\partial U_n}{\partial t} Y_j \, dS = \rho_0 \frac{dU_i}{dt} \oint_S (n_i y_j - n_i \varphi_j^*) \, dS$$

$$= (m_0 \delta_{ij} + M_{ij}) \frac{dU_i}{dt},$$

where if the body has volume Δ, then $m_0 = \rho_0 \Delta$ is the *mass of fluid* displaced by the body, and M_{ij} is the added mass tensor of the body (see (3.8.6)).

The solution (5.4.1) can therefore be written

$$p(\mathbf{x}, t) \approx \frac{-x_j}{4\pi c_0 |\mathbf{x}|^2} \frac{\partial}{\partial t} \left[\rho_0 \int (\boldsymbol{\omega} \wedge \mathbf{v}) \cdot \nabla Y_j \, d^3 \mathbf{y} - \eta \oint_S \boldsymbol{\omega} \wedge \nabla Y_j \cdot d\mathbf{S}(\mathbf{y}) \right.$$

$$\left. - (m_0 \delta_{ij} + M_{ij}) \frac{dU_i}{dt} \right]. \tag{5.4.2}$$

Reference to Equation (4.4.4) shows that this can also be written

$$p(\mathbf{x}, t) \approx \frac{x_j}{4\pi c_0 |\mathbf{x}|^2} \frac{\partial F_j}{\partial t} \left(t - \frac{|\mathbf{x}|}{c_0} \right) + \frac{m_0 x_j}{4\pi c_0 |\mathbf{x}|^2} \frac{\partial^2 U_j}{\partial t^2} \left(t - \frac{|\mathbf{x}|}{c_0} \right), \quad |\mathbf{x}| \to \infty, \tag{5.4.3}$$

where $\mathbf{F}(t)$ is the unsteady force exerted on the fluid by the body. This is just our earlier conclusion (2.4.2) derived from Curle's equation, with the addition of the fluid-displacement effect of the vibrating body, and is equivalent to the solution (3.8.10) obtained in the absence of vorticity.

The relative contributions from the volume and surface distributions of vorticity in (5.4.2) (respectively the first and second integrals) for turbulence of length scale ℓ and velocity v are estimated respectively by

$$\rho_0 v^2 M \frac{\ell}{|\mathbf{x}|} \quad \text{and} \quad \rho_0 v^2 M \frac{\ell}{|\mathbf{x}|} \frac{1}{\text{Re}}, \quad \text{Re} = \frac{v\ell}{\nu}.$$

Thus, in high Reynolds number turbulent flows the surface frictional contribution to the dipole force $\mathbf{F}$ can usually be neglected. For a *nonvibrating* compact body the principal component of the acoustic pressure in the far field is therefore given by

$$p(\mathbf{x}, t) \approx \frac{-\rho_0 x_j}{4\pi c_0 |\mathbf{x}|^2} \frac{\partial}{\partial t} \int (\boldsymbol{\omega} \wedge \mathbf{v}) \left(\mathbf{y}, t - \frac{|\mathbf{x}|}{c_0}\right) \cdot \nabla Y_j(\mathbf{y}) \, d^3 \mathbf{y}, \quad |\mathbf{x}| \to \infty.$$

(5.4.4)

5.5 Radiation from Cylindrical Bodies of Compact Cross Section

An important special case occurs when vorticity interacts with a cylindrical (or approximately cylindrical) surface S of compact cross section (such as the strip airfoil of Fig. 3.6.3). If the vorticity extends over an extensive spanwise section of the body it may be important to account for differences in the retarded times of the sound produced at different spanwise positions.

To do this we first write the Kirchhoff vector in the form

$$\mathbf{Y} = \mathbf{Y}_\perp + \mathbf{k} y_3, \quad \mathbf{Y}_\perp = (Y_1(\mathbf{y}), Y_2(\mathbf{y}), 0), \tag{5.5.1}$$

where $\mathbf{k}$ is a unit vector in the x_3 direction. Then, because variations in the spanwise source position are not necessarily small compared to the acoustic wavelength, the compact Green's function in (5.3.6) is expanded as follows:

$$
\begin{aligned}
G(\mathbf{x}, \mathbf{y}, t - \tau) &= \frac{1}{4\pi |\mathbf{X} - \mathbf{Y}|} \delta\left(t - \tau - \frac{|\mathbf{X} - \mathbf{Y}|}{c_0}\right) \\
&\approx \frac{1}{4\pi |\mathbf{x} - \mathbf{k} y_3|} \delta\left(t - \tau - \frac{|\mathbf{x} - \mathbf{k} y_3|}{c_0}\right) \\
&\quad + \frac{x_j Y_{\perp j}}{4\pi c_0 |\mathbf{x} - \mathbf{k} y_3|^2} \delta'\left(t - \tau - \frac{|\mathbf{x} - \mathbf{k} y_3|}{c_0}\right) \\
&\approx \frac{1}{4\pi |\mathbf{x}|} \delta\left(t - \tau - \frac{|\mathbf{x} - \mathbf{k} y_3|}{c_0}\right) \\
&\quad + \frac{x_j Y_{\perp j}}{4\pi c_0 |\mathbf{x}|^2} \delta'\left(t - \tau - \frac{|\mathbf{x} - \mathbf{k} y_3|}{c_0}\right), \quad |\mathbf{x}| \to \infty. \quad (5.5.2)
\end{aligned}
$$

For example, when a stationary cylindrical body interacts with high Reynolds number flow at low Mach number, the monopole term in (5.5.2) can be discarded as before, and the acoustic pressure given by (5.3.6) becomes

$$p(\mathbf{x}, t) \approx \frac{-\rho_0 x_j}{4\pi c_0 |\mathbf{x}|^2} \frac{\partial}{\partial t} \int (\boldsymbol{\omega} \wedge \mathbf{v}) \left(\mathbf{y}, t - \frac{|\mathbf{x} - \mathbf{k} y_3|}{c_0}\right) \cdot \nabla Y_{\perp j}(\mathbf{y}) \, d^3 \mathbf{y}, \quad |\mathbf{x}| \to \infty.$$

(5.5.3)

The sound is produced by dipole sources orientated in the lift and drag directions only ($j = 1, 2$). This approximation is applicable also to a thin airfoil of large but finite span, and in cases where the chord of the airfoil is a slowly varying function of y_3 (such as the elliptic airfoil of Fig. 3.9.1).

5.6 Impulse Theory of Vortex Sound

An interesting formula for the sound generated by vorticity near a compact body can be derived directly from the representation (4.3.3) of the velocity $\mathbf{v}(\mathbf{x}, t)$ in the *hydrodynamic* far field in terms of the *impulse* $\mathbf{I}(t)$. Indeed,

$$\mathbf{v}(\mathbf{x}, t) \approx \nabla \varphi \quad \text{when } |\mathbf{x}| \gg \ell,$$

where ℓ is the length scale of the interaction region (the body), and

$$\varphi(\mathbf{x}, t) = \operatorname{div}\left(\frac{\mathbf{I}(t)}{4\pi|\mathbf{x}|}\right) \quad \text{where } \mathbf{I}(t) = \frac{1}{2}\int \mathbf{y} \wedge \boldsymbol{\omega}(\mathbf{y}, t) \, d^3\mathbf{y}. \quad (5.6.1)$$

This expression for $\varphi(\mathbf{x}, t)$ defines the incompressible motion in the irrotational region far from the body. It is the velocity potential of a hydrodynamic dipole that will be recognized as the *acoustic near field* of an outgoing acoustic dipole representing sound production by the flow (see Equations (1.7.4) and (1.7.9)). The acoustic dipole is found simply by replacing $\mathbf{I}(t)$ in (5.6.1) by $\mathbf{I}(t - |\mathbf{x}|/c_0)$. At large distances from the body (where the undisturbed fluid is stationary) the pressure $p(\mathbf{x}, t) = -\rho_0 \partial\varphi/\partial t$, and this procedure therefore leads to the following formula for the sound in terms of the vorticity:

$$p(\mathbf{x}, t) \approx -\rho_0 \frac{\partial}{\partial x_j}\left(\frac{1}{4\pi|\mathbf{x}|}\frac{\partial I_j}{\partial t}(t - |\mathbf{x}|/c_0)\right)$$

$$\approx \frac{\rho_0 x_j}{4\pi c_0 |\mathbf{x}|^2}\frac{\partial^2 I_j}{\partial t^2}\left(t - \frac{|\mathbf{x}|}{c_0}\right), \quad |\mathbf{x}| \to \infty$$

$$= \frac{\rho_0 x_j}{8\pi c_0 |\mathbf{x}|^2}\frac{\partial^2}{\partial t^2}\int (\mathbf{y} \wedge \boldsymbol{\omega})_j\left(\mathbf{y}, t - \frac{|\mathbf{x}|}{c_0}\right) d^3\mathbf{y}. \quad (5.6.2)$$

Equation (4.4.1) shows that this is equivalent to the compact approximation (5.4.3). Note, however, that $\boldsymbol{\omega}$ in (5.6.2) is the generalized vorticity, including the bound vorticity on S. This should be contrasted with the representation (5.4.2) involving the Kirchhoff vector, where bound vorticity occurs only in the surface integral of the frictional contribution to the sound. But, (5.4.2) is valid only for a body in *translational* motion whereas (5.6.2) is applicable for a body executing any combination of translations and rotations (Fig. 4.1.1).

The impulse $\mathbf{I}$ is *constant* for vorticity in an unbounded fluid (when compressibility is ignored). We know that the sound is now generated by quadrupole sources and that it can be represented in terms of the vorticity as in (5.1.6). Möhring (1978) has shown that it is also possible to express the quadrupole sound as a third-order time derivative of a *second-order moment* of the vorticity, analogous to the first order moment in (5.6.2) (see Problems 5), namely

$$p(\mathbf{x}, t) \approx \frac{\rho_0 x_i x_j}{12\pi c_0^2 |\mathbf{x}|^3} \frac{\partial^3}{\partial t^3} \int y_i (\mathbf{y} \wedge \boldsymbol{\omega})_j (\mathbf{y}, t - |\mathbf{x}|/c_0) \, d^3\mathbf{y}, \quad |\mathbf{x}| \to \infty. \quad (5.6.3)$$

Problems 5

1. **Kirchhoff's spinning vortex:** A columnar vortex parallel to the x_3 axis has elliptic cross section defined by the polar equation $r = a\{1 + \epsilon \cos(2\theta - \Omega t/2)\}$, where $\epsilon \ll 1$ and Ω is the uniform vorticity in the core. The ellipse rotates at angular velocity $\frac{1}{4}\Omega$, and the velocity distribution within the core is given by

$$\mathbf{v} = (v_1, v_2) = -\frac{1}{2}\Omega r(\sin\theta + \epsilon \sin(\theta - \Omega t/2), -\cos\theta + \epsilon \cos(\theta - \Omega t/2)).$$

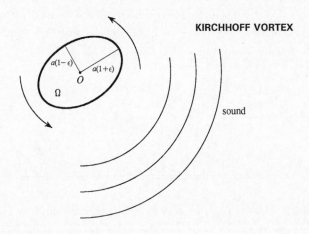

Show that to first order in ϵ the vortex is equivalent to the two-dimensional quadrupole

$$\text{div}(\boldsymbol{\omega} \wedge \mathbf{v}) \approx \frac{\partial^2}{\partial x_i \partial x_j}(T_{ij}\delta(x_1)\delta(x_2)), \quad i, j = 1, 2$$

where

$$T_{ij} = \frac{\epsilon \pi \Omega^2 a^4}{8} \begin{pmatrix} \cos(\Omega t/2) & \sin(\Omega t/2) \\ \sin(\Omega t/2) & -\cos(\Omega t/2) \end{pmatrix}$$

and that the acoustic pressure is

$$p \approx -\frac{\epsilon}{8}\sqrt{\frac{2\pi a}{r}}\rho_0 U^2 M^{3/2} \cos\left[2\theta - \frac{\Omega}{2}\left(t - \frac{r}{c_0}\right) + \frac{\pi}{4}\right], \qquad \frac{\Omega r}{c_0} \to \infty,$$

where $U = \frac{1}{2}a\Omega$ and $M = U/c_0$.

2. **Coaxial vortex rings:** Use Equation (5.2.7) to calculate the sound produced by the unsteady motions of an acoustically compact system of N vortex rings coaxial with the x_1 axis. Take the vorticity of the nth vortex to be $\omega_n = \Gamma_n \delta(x_1 - X_n(t))\delta(r - R_n(t))\mathbf{i}_\theta$, where (r, θ, x_1) are cylindrical polar coordinates, $R_n(t)$ being the vortex ring radius, $X_n(t)$ its location in the x_1 direction, and $\mathbf{i}_\theta$ is a unit vector in the azimuthal direction. Show that

$$p \approx \frac{\rho_0}{8c_0^2|\mathbf{x}|}(3\cos^2 \Theta - 1)\frac{\partial^2}{\partial t^2}\left[\sum_n \Gamma_n X_n \frac{dR_n^2}{dt}\right], \qquad |\mathbf{x}| \to \infty,$$

where Θ is the angle between the observer direction and the positive x_1 axis and the term in square braces is evaluated at $\tau = t - |\mathbf{x}|/c_0$.

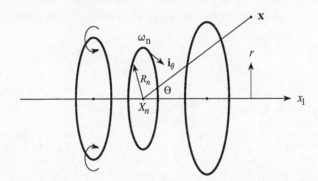

In an ideal, incompressible fluid conservation of energy and momentum implies that

$$\sum_n \Gamma_n R_n \left(R_n \frac{dX_n}{dt} - X_n \frac{dR_n}{dt}\right) = \text{constant}, \qquad \sum_n \Gamma_n R_n^2 = \text{constant}.$$

Use these equations to show that

$$p \approx \frac{\rho_0}{12c_0^2|\mathbf{x}|}(3\cos^2 \Theta - 1)\left[\frac{d^3 S}{dt^3}\right], \qquad |\mathbf{x}| \to \infty,$$

where $S(t) = \sum_n \Gamma_n R_n^2 X_n$.

3. Calculate the sound produced by a vortex ring of total circulation Γ, coaxial with the x_1 axis, whose core has elliptic cross section of major and minor axes $2a, 2b \ll R$, where R is the mean radius of the ring. Assume that the x_1 coordinate $X(t)$ of the vorticity centroid satisfies

$$\frac{dX}{dt} = \frac{\Gamma}{4\pi R}\left[\ln\left(\frac{16R}{a+b}\right) - \frac{1}{4} + \frac{3(a-b)}{2(a+b)}\cos 2\Omega t\right], \quad \Omega = \frac{\Gamma}{\pi(a+b)^2}.$$

In the notation of Question 2, show that

$$p \approx \frac{\rho_0 \Gamma R^2 (3\cos^2\Theta - 1)}{12 c_0^2 |\mathbf{x}|}\left[\frac{d^3 X}{dt^3}\right]_{t-\frac{|\mathbf{x}|}{c_0}}, \quad |\mathbf{x}| \to \infty$$

$$= \frac{\rho_0 U^2 M^2}{8\pi^3}\frac{R(3\cos^2\Theta - 1)}{|\mathbf{x}|}\left(\frac{a-b}{a+b}\right)\cos\left(\frac{2\Gamma}{\pi(a+b)^2}\left[t - \frac{|\mathbf{x}|}{c_0}\right]\right),$$

$$\text{where } U = \frac{\Gamma}{a+b}, \quad M = \frac{U}{c_0}.$$

4. Derive Möhring's formula (5.6.3) for the acoustic pressure generated by a compact region of vorticity in an unbounded fluid. Take the cross product of $\mathbf{y}$ with the high Reynolds number vorticity equation $\partial\boldsymbol{\omega}/\partial t + \text{curl}\,(\boldsymbol{\omega} \wedge \mathbf{v}) = 0$ (expressed in terms of $\mathbf{y}$ and t as independent variables), multiply by y_i, and use the identity

$$\mathbf{y} \wedge \text{curl}\,\mathbf{A} = 2\mathbf{A} + \nabla(\mathbf{y}\cdot\mathbf{A}) - \frac{\partial}{\partial y_j}(y_j\mathbf{A})$$

to deduce that the integral in (5.1.6) can be written

$$\int y_i(\boldsymbol{\omega} \wedge \mathbf{v})_j\, d^3\mathbf{y} = -\frac{1}{3}\frac{\partial}{\partial t}\int y_i(\mathbf{y} \wedge \boldsymbol{\omega})_j\, d^3\mathbf{y} + \frac{1}{3}\delta_{ij}\int \frac{1}{2}v^2\, d^3\mathbf{y}.$$

The result now follows by noting that the estimate (5.1.7) permits the second integral on the right to be discarded.

5. The free space Green's function in two dimensions – the solution of

$$\left(\frac{1}{c_0^2}\frac{\partial^2}{\partial t^2} - \nabla^2\right)G = \delta(x_1 - y_1)\delta(x_2 - y_2)\delta(t - \tau), \quad \text{where } G = 0 \quad \text{for } t < \tau,$$

– can be derived by integrating the three-dimensional Green's function

$$\frac{1}{4\pi|\mathbf{x} - \mathbf{y}|}\delta\left(t - \tau - \frac{|\mathbf{x} - \mathbf{y}|}{c_0}\right) \quad \text{over } -\infty < y_3 < \infty.$$

Deduce that

$$G(\mathbf{x}, \mathbf{y}, t - \tau) = \frac{H(t - \tau - |\mathbf{x} - \mathbf{y}|/c_0)}{2\pi\sqrt{(t - \tau)^2 - (\mathbf{x} - \mathbf{y})^2/c_0^2}}, \quad \mathbf{x} = (x_1, x_2), \quad \mathbf{y} = (y_1, y_2),$$

and that near the wavefront, where $|\mathbf{x} - \mathbf{y}| \approx c_0(t - \tau)$

$$G(\mathbf{x}, \mathbf{y}, t - \tau) \approx \frac{H(t - \tau - |\mathbf{x} - \mathbf{y}|/c_0)}{2\pi\sqrt{2|\mathbf{x} - \mathbf{y}|/c_0}\sqrt{(t - \tau) - |\mathbf{x} - \mathbf{y}|/c_0}}.$$

6. Consider sound production by a compact distribution of vorticity in an unbounded two-dimensional flow (independent of x_3), where the vorticity $\boldsymbol{\omega}$ is parallel to the $\mathbf{k}$ direction (the x_3 axis). Show that

$$\int y_i(\boldsymbol{\omega} \wedge \mathbf{v})_j \, dy_1 \, dy_2 \approx -\frac{1}{2}\frac{\partial}{\partial t} \int y_i(\mathbf{y} \wedge \boldsymbol{\omega})_j \, dy_1 \, dy_2, \quad \text{where } i, j = 1, 2.$$

Deduce Möhring's (1980) two-dimensional representation

$$p(\mathbf{x}, t) \approx \frac{\rho_0 x_i x_j}{4\pi c_0^2 |\mathbf{x}|^2}\frac{\partial^3}{\partial t^3} \int dy_1 \, dy_2 \int_{-\infty}^{t - |\mathbf{x}|/c_0} \frac{y_i(\mathbf{y} \wedge \boldsymbol{\omega})_j(\mathbf{y}, \tau) \, d\tau}{\sqrt{(t - \tau)^2 - |\mathbf{x}|^2/c_0^2}}$$

$$\approx \frac{\rho_0 x_i x_j}{4\pi c_0^2 |\mathbf{x}|^2}\sqrt{\frac{c_0}{2|\mathbf{x}|}}\frac{\partial^3}{\partial t^3} \int dy_1 \, dy_2 \int_{-\infty}^{t - |\mathbf{x}|/c_0} \frac{y_i(\mathbf{y} \wedge \boldsymbol{\omega})_j(\mathbf{y}, \tau) \, d\tau}{\sqrt{t - \tau - |\mathbf{x}|/c_0}},$$

$$|\mathbf{x}| \to \infty.$$

7. Use the result of Problem 6 to derive the Solution (5.2.10) for the sound produced by a spinning vortex pair.
8. Use the result of Problem 6 to solve Problem 1.

6

Vortex–Surface Interaction Noise
in Two Dimensions

6.1 Compact Green's Function in Two Dimensions

In this chapter, we apply the general high Reynolds number solution (5.3.6) to determine sound produced by two-dimensional interactions of rectilinear vortices with a stationary solid boundary. Conditions are assumed to be uniform in the x_3 direction, parallel to the vorticity. We shall derive the two-dimensional analogue of the general solution

$$\frac{p}{\rho_0}(\mathbf{x}, t) = -\int_V (\boldsymbol{\omega} \wedge \mathbf{v})(\mathbf{y}, \tau) \cdot \frac{\partial G}{\partial \mathbf{y}}(\mathbf{x}, \mathbf{y}, t - \tau) \, d^3\mathbf{y} \, d\tau \qquad (6.1.1)$$

by first determining a suitable representation of G in two dimensions. Both the vorticity convection velocity $\mathbf{v}$ and the Lamb vector $\boldsymbol{\omega} \wedge \mathbf{v}$ are parallel to the $x_1 x_2$ plane, so that only the y_1 and y_2 components of the gradient $\partial G / \partial \mathbf{y}$ contribute to the integral. Also, because $(\boldsymbol{\omega} \wedge \mathbf{v})(\mathbf{y}, \tau)$ depends only on y_1, y_2 and τ, the integration with respect to the spanwise coordinate y_3 involves only the Green's function, and may be performed prior to any further calculations of the sound.

Let

$$G_2 = \int_{-\infty}^{\infty} G(\mathbf{x}, \mathbf{y}, t - \tau) \, dy_3. \qquad (6.1.2)$$

For two-dimensional problems G is a function of $y_3 - x_3$, and the function G_2 will therefore satisfy the Green's function equation

$$\left(\frac{1}{c_0^2} \frac{\partial^2}{\partial t^2} - \nabla^2 \right) G_2 = \delta(x_1 - y_1)\delta(x_2 - y_2)\delta(t - \tau),$$

$$\text{where } G_2 = 0 \quad \text{for } t < \tau, \qquad (6.1.3)$$

obtained by integrating the three-dimensional Equation (3.1.4) over $-\infty < y_3 < \infty$. In two dimensions, G_2 represents the field generated by a uniform *line* source parallel to the x_3 axis extending along the whole of the line $x_1 = y_1$, $x_2 = y_2$.

Similarly, the compact Green's function for a cylindrical body (with generators parallel to x_3)

$$G(\mathbf{x}, \mathbf{y}, t - \tau) = \frac{1}{4\pi |\mathbf{X} - \mathbf{Y}|} \delta \left(t - \tau - \frac{|\mathbf{X} - \mathbf{Y}|}{c_0} \right),$$

$$X_{1,2} = x_{1,2} - \varphi_{1,2}^*(\mathbf{x}), \quad Y_{1,2} = y_{1,2} - \varphi_{1,2}^*(\mathbf{y}), \quad X_3 = x_3, \quad Y_3 = y_3,$$

is a function of $y_3 - x_3$, and the corresponding two-dimensional *compact* Green's function can be found by integrating over $-\infty < y_3 < \infty$.

Set $\xi = y_3 - x_3$, $\bar{\mathbf{x}} = (x_1, x_2)$ and let $|\bar{\mathbf{x}}| = (x_1^2 + x_2^2)^{\frac{1}{2}} \to \infty$. Taking the origin of coordinates within the cylindrical body, we have

$$G_2 \approx \frac{1}{4\pi} \int_{-\infty}^{\infty} \delta \left(t - \tau - \frac{\sqrt{|\bar{\mathbf{x}}|^2 + \xi^2}}{c_0} + \frac{\bar{\mathbf{x}} \cdot \mathbf{Y}}{c_0 \sqrt{|\bar{\mathbf{x}}|^2 + \xi^2}} \right) \frac{d\xi}{\sqrt{|\bar{\mathbf{x}}|^2 + \xi^2}}$$

$$\text{as } |\bar{\mathbf{x}}| \to \infty,$$

where $\bar{\mathbf{x}} \cdot \mathbf{Y} = x_1 Y_1 + x_2 Y_2$ is independent of ξ. To use this to evaluate (6.1.1) the δ function must be expanded to first order in $\mathbf{Y}$

$$G_2 \approx \frac{1}{4\pi} \int_{-\infty}^{\infty} \delta \left(t - \tau - \frac{\sqrt{|\bar{\mathbf{x}}|^2 + \xi^2}}{c_0} \right) \frac{d\xi}{\sqrt{|\bar{\mathbf{x}}|^2 + \xi^2}}$$

$$+ \frac{\bar{\mathbf{x}} \cdot \mathbf{Y}}{4\pi c_0} \int_{-\infty}^{\infty} \delta' \left(t - \tau - \frac{\sqrt{|\bar{\mathbf{x}}|^2 + \xi^2}}{c_0} \right) \frac{d\xi}{(|\bar{\mathbf{x}}|^2 + \xi^2)}.$$

Only the second term on the right depends on $\mathbf{y}$ and therefore contributes to the radiation integral (6.1.1). We can therefore take

$$G_2 \approx \frac{\bar{\mathbf{x}} \cdot \mathbf{Y}}{4\pi c_0} \int_{-\infty}^{\infty} \delta' \left(t - \tau - \frac{\sqrt{|\bar{\mathbf{x}}|^2 + \xi^2}}{c_0} \right) \frac{d\xi}{(|\bar{\mathbf{x}}|^2 + \xi^2)}$$

$$= \frac{\bar{\mathbf{x}} \cdot \mathbf{Y}}{2\pi c_0} \frac{\partial}{\partial t} \int_0^{\infty} \delta \left(t - \tau - \frac{\sqrt{|\bar{\mathbf{x}}|^2 + \xi^2}}{c_0} \right) \frac{d\xi}{(|\bar{\mathbf{x}}|^2 + \xi^2)}$$

$$= \frac{\bar{\mathbf{x}} \cdot \mathbf{Y}}{2\pi c_0} \frac{\partial}{\partial t} \left\{ \frac{H(t - \tau - |\bar{\mathbf{x}}|/c_0)}{(|\bar{\mathbf{x}}|^2 + \xi^2) \left| \frac{\partial}{\partial \xi} \frac{\sqrt{|\bar{\mathbf{x}}|^2 + \xi^2}}{c_0} \right|} \right\}_{\xi = \sqrt{c_0^2 (t - \tau)^2 - |\bar{\mathbf{x}}|^2}}$$

$$= \frac{\bar{\mathbf{x}} \cdot \mathbf{Y}}{2\pi c_0} \frac{\partial}{\partial t} \left\{ \frac{H(t - \tau - |\bar{\mathbf{x}}|/c_0)}{(t - \tau) \sqrt{c_0^2 (t - \tau)^2 - |\bar{\mathbf{x}}|^2}} \right\}.$$

We shall henceforth in this chapter regard all space vectors as two-dimensional, such as $\mathbf{x} = (x_1, x_2)$, $\mathbf{y} = (y_1, y_2)$, and drop the overbar on $\bar{\mathbf{x}}$ and the subscript 2 on G_2 and soforth. The dipole component of the two-dimensional compact Green's function then becomes

$$G(\mathbf{x}, \mathbf{y}, t - \tau) \approx \frac{\mathbf{x} \cdot \mathbf{Y}}{2\pi c_0} \frac{\partial}{\partial t} \left\{ \frac{H(t - \tau - |\mathbf{x}|/c_0)}{(t - \tau)\sqrt{c_0^2(t - \tau)^2 - |\mathbf{x}|^2}} \right\}, \quad |\mathbf{x}| \to \infty.$$

(6.1.4)

In contrast to the three-dimensional Green's function, which is nonzero only on a spherically expanding wavefront, G has an infinite peak at the wavefront where $t - \tau - |\mathbf{x}|/c_0 = 0$ followed by a slowly decaying tail. At any point $\mathbf{x}$ in the far field the first sound arrives from the nearest point on the line source of (6.1.3), after propagating along a ray perpendicular to the source over a distance equal to $|\mathbf{x}|$, and therefore after a time delay $|\mathbf{x}|/c_0$; but the observer at $\mathbf{x}$ receives sound continuously after the passage of this wavefront, generated at more distant sections of the infinitely long line source; the wavefront arrives at time $\tau + |\mathbf{x}|/c_0$, and at a later time t sound is received from those source points whose distance from $\mathbf{x}$ is equal to $c_0(t - \tau)$.

The far-field representation (6.1.4) can be approximated further by expanding about the wavefront (where $t - \tau = |\mathbf{x}|/c_0$), which contains most of the acoustic energy. Just to the rear of the wavefront

$$(t - \tau)\sqrt{c_0^2(t - \tau)^2 - |\mathbf{x}|^2} \equiv (t - \tau)\sqrt{c_0(t - \tau) + |\mathbf{x}|}\sqrt{c_0(t - \tau) - |\mathbf{x}|}$$

$$\approx \frac{|\mathbf{x}|}{c_0}\sqrt{2|\mathbf{x}|}\sqrt{c_0(t - \tau) - |\mathbf{x}|}, \quad \text{for } t - \tau \sim \frac{|\mathbf{x}|}{c_0}.$$

Therefore, (6.1.4) becomes

$$G(\mathbf{x}, \mathbf{y}, t - \tau) \approx \frac{\mathbf{x} \cdot \mathbf{Y}}{2\pi \sqrt{2c_0}|\mathbf{x}|^{\frac{3}{2}}} \frac{\partial}{\partial t} \left\{ \frac{H(t - \tau - |\mathbf{x}|/c_0)}{\sqrt{t - \tau - |\mathbf{x}|/c_0}} \right\}, \quad |\mathbf{x}| \to \infty, \quad (6.1.5)$$

(see Question 1 of Problems 6).

The special case of a cylindrical body adjacent to a plane, rigid wall at $x_2 = 0$, or of a cylindrical wall cavity or projection from a wall (see Fig. 3.9.2) is handled by the two-dimensional version of the compact Green's function (3.9.3). The procedure described above yields the following expression for the dipole component of the two-dimensional compact Green's function

$$G(\mathbf{x}, \mathbf{y}, t - \tau) \approx \frac{x_1 Y_1}{\pi \sqrt{2c_0}|\mathbf{x}|^{\frac{3}{2}}} \frac{\partial}{\partial t} \left\{ \frac{H(t - \tau - |\mathbf{x}|/c_0)}{\sqrt{t - \tau - |\mathbf{x}|/c_0}} \right\}, \quad |\mathbf{x}| \to \infty, \quad (6.1.6)$$

where $Y_1 \equiv Y_1(y_1, y_2)$ is the velocity potential of incompressible flow past the cylinder in a direction parallel to the wall (with unit speed at large distances from the body). G represents the field of a dipole orientated parallel to the wall and perpendicular to the cylinder axis, and the effect of the wall is to generate an equal image dipole that just *doubles* the magnitude of the sound relative to the corresponding dipole (6.1.5) of the cylinder in the absence of the wall.

6.2 Sound Generated by a Line Vortex Interacting with a Cylindrical Body

The sound produced by two-dimensional vortex motion at low Mach number $M \sim v/c_0$ near a stationary, rigid cylindrical body of diameter ℓ has characteristic wavelength $\sim \ell/M \gg \ell$. The acoustic pressure is determined by the two-dimensional version of (6.1.1) using the compact Green's function (6.1.5), viz,

$$
p(\mathbf{x}, t) \approx \frac{-\rho_0 x_j}{2\pi \sqrt{2c_0} |\mathbf{x}|^{\frac{3}{2}}} \frac{\partial}{\partial t} \int_{-\infty}^{t-|\mathbf{x}|/c_0} \frac{d\tau}{\sqrt{t - \tau - |\mathbf{x}|/c_0}}
$$
$$
\times \int (\boldsymbol{\omega} \wedge \mathbf{v} \cdot \nabla Y_j)(\mathbf{y}, \tau) \, dy_1 \, dy_2, \quad |\mathbf{x}| \to \infty. \quad (6.2.1)
$$

According to the inviscid form of the Formula (4.4.4) applied to a stationary body, this can also be written

$$
p(\mathbf{x}, t) \approx \frac{x_j}{2\pi \sqrt{2c_0} |\mathbf{x}|^{\frac{3}{2}}} \frac{\partial}{\partial t} \int_{-\infty}^{t-|\mathbf{x}|/c_0} \frac{F_j(\tau) \, d\tau}{\sqrt{t - \tau - |\mathbf{x}|/c_0}}, \quad |\mathbf{x}| \to \infty, \quad (6.2.2)
$$

where F_j is the force *per unit length of the cylinder* exerted on the fluid in the j direction.

In these two-dimensional problems the vorticity $\boldsymbol{\omega}$ is directed along the x_3 axis, out of the plane of the paper in Fig. 6.2.1 and parallel to the generators of the cylinder. Let $\mathbf{k}$ be a unit vector in this direction and consider a line vortex of strength Γ whose position and translational velocity are

$$
\mathbf{x} = \mathbf{x}_0(t), \qquad \mathbf{v} = \frac{d\mathbf{x}_0}{dt}(t).
$$

If the motion elsewhere is irrotational, we have

$$
\boldsymbol{\omega} = \Gamma \mathbf{k} \delta(\mathbf{x} - \mathbf{x}_0(t))
$$

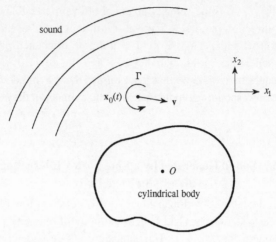

Fig. 6.2.1.

so that

$$\boldsymbol{\omega} \wedge \mathbf{v} = \Gamma \mathbf{k} \wedge \mathbf{v}\delta(\mathbf{x} - \mathbf{x}_0(t)) \equiv \Gamma \mathbf{k} \wedge \frac{d\mathbf{x}_0}{dt}(t)\delta(\mathbf{x} - \mathbf{x}_0(t)), \quad (6.2.3)$$

and (6.2.1) yields the following general formula for the acoustic pressure as $|\mathbf{x}| \to \infty$:

$$
\begin{aligned}
p(\mathbf{x}, t) \\
\approx \frac{-\rho_0 \Gamma x_j}{2\pi\sqrt{2c_0}|\mathbf{x}|^{\frac{3}{2}}} \frac{\partial}{\partial t} \int_{-\infty}^{t-|\mathbf{x}|/c_0} &\left(\mathbf{k} \wedge \frac{d\mathbf{x}_0}{d\tau}(\tau) \cdot \nabla Y_j(\mathbf{x}_0(\tau)) \right) \frac{d\tau}{\sqrt{t - \tau - |\mathbf{x}|/c_0}} \\
= \frac{-\rho_0 \Gamma x_j}{2\pi\sqrt{2c_0}|\mathbf{x}|^{\frac{3}{2}}} \frac{\partial}{\partial t} \int_{-\infty}^{t-|\mathbf{x}|/c_0} &\left\{ \frac{dx_{01}}{d\tau} \frac{\partial Y_j}{\partial y_2} - \frac{dx_{02}}{d\tau} \frac{\partial Y_j}{\partial y_1} \right\}_{\mathbf{x}_0(\tau)} \frac{d\tau}{\sqrt{t - \tau - |\mathbf{x}|/c_0}}.
\end{aligned}
$$
$$(6.2.4)$$

6.2.1 Example 1: Sound Produced by Vortex Motion near a Circular Cylinder

Assume there is no mean flow, and let the vortex strength Γ be sufficiently small for the local motion to be considered incompressible. Then, the vortex will traverse the circular orbit discussed in Section 4.6 (Example 2).

Let the cylinder have radius a, the vortex path have radius r_0, and take the coordinate origin at the centre of the cylinder. If $\Gamma > 0$ the vortex moves in the

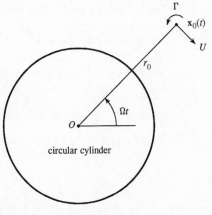

Fig. 6.2.2.

clockwise direction in Fig. 6.2.2 at the speed given by (4.6.8) when r is replaced by r_0. Then,

$$\mathbf{v} = \frac{d\mathbf{x}_0}{dt}(t) = \Omega \mathbf{k} \wedge \mathbf{x}_0(t),$$

$$\Omega = \frac{-\Gamma a^2}{2\pi r_0^2 (r_0^2 - a^2)},$$

$$\mathbf{x}_0 = r_0(\cos \Omega t, \sin \Omega t),$$

and

$$\mathbf{k} \wedge \frac{d\mathbf{x}_0}{dt} = \Omega \mathbf{k} \wedge (\mathbf{k} \wedge \mathbf{x}_0) = -\Omega \mathbf{x}_0.$$

Therefore, (6.2.4) becomes

$$p(\mathbf{x}, t) \approx \frac{\rho_0 \Gamma \Omega x_j r_0}{2\pi \sqrt{2c_0} |\mathbf{x}|^{\frac{3}{2}}} \frac{\partial}{\partial t} \int_{-\infty}^{t-|\mathbf{x}|/c_0} \frac{\partial Y_j}{\partial r}(\mathbf{x}_0(\tau)) \frac{d\tau}{\sqrt{t - \tau - |\mathbf{x}|/c_0}}, \quad (6.2.5)$$

where $r = \sqrt{y_1^2 + y_2^2}$ is the radial distance from the cylinder axis.

Table 3.9.1 supplies the components of the two-dimensional Kirchhoff vector for the cylinder:

$$Y_1 = \cos \vartheta \left(r + \frac{a^2}{r} \right), \qquad Y_2 = \sin \vartheta \left(r + \frac{a^2}{r} \right),$$

where $(y_1, y_2) = r(\cos \vartheta, \sin \vartheta)$. Thus, by introducing polar coordinates for $\mathbf{x}$

$$\mathbf{x} = |\mathbf{x}|(\cos \theta, \sin \theta)$$

we find $\quad x_j \dfrac{\partial Y_j}{\partial r} = |\mathbf{x}| \left(1 - \dfrac{a^2}{r^2} \right) \cos(\theta - \vartheta)$

so that

$$x_j \frac{\partial Y_j}{\partial r}(\mathbf{x}_0(\tau)) = |\mathbf{x}| \left(1 - \frac{a^2}{r_0^2} \right) \cos(\theta - \Omega \tau),$$

and (6.2.5) becomes

$$p(\mathbf{x}, t) \approx \frac{\rho_0 \Gamma \Omega r_0}{2\pi (2c_0 |\mathbf{x}|)^{\frac{1}{2}}} \left(1 - \frac{a^2}{r_0^2} \right) \frac{\partial}{\partial t} \int_{-\infty}^{t-|\mathbf{x}|/c_0} \frac{\cos(\theta - \Omega \tau) d\tau}{\sqrt{t - \tau - |\mathbf{x}|/c_0}}.$$

Hence, using the formula

$$\int_{-\infty}^{t-|\mathbf{x}|/c_0} \frac{\cos(\theta - \Omega \tau) d\tau}{\sqrt{t - \tau - |\mathbf{x}|/c_0}} = \left(\frac{\pi}{|\Omega|} \right)^{\frac{1}{2}} \cos \left\{ \theta - \Omega \left[t - \frac{|\mathbf{x}|}{c_0} \right] - \frac{\pi}{4} \right\}$$

the pressure becomes

$$p(\mathbf{x}, t) \approx \frac{\rho_0 \Gamma |\Omega|^{\frac{3}{2}} r_0}{2(2\pi c_0 |\mathbf{x}|)^{\frac{1}{2}}} \left(1 - \frac{a^2}{r_0^2} \right) \sin \left\{ \theta - \Omega \left[t - \frac{|\mathbf{x}|}{c_0} \right] - \frac{\pi}{4} \right\}$$

$$= \rho_0 U^2 \sqrt{M} \sqrt{\frac{\pi r_0}{2|\mathbf{x}|}} \left(\frac{r_0}{a} \right)^2 \left(1 - \frac{a^2}{r_0^2} \right)^2 \sin \left\{ \theta - \Omega \left[t - \frac{|\mathbf{x}|}{c_0} \right] - \frac{\pi}{4} \right\},$$

$$|\mathbf{x}| \to \infty, \quad (6.2.6)$$

where $U = |\Omega| r_0$ is the vortex speed.

The acoustic waves decay like $1/\sqrt{|\mathbf{x}|}$ with distance, which is appropriate for energy spreading two dimensionally in *cylindrically diverging waves*, and have the characteristic dipole amplitude proportional to $\rho_0 U^2 \sqrt{M}$. In three dimensions the pressure would be proportional to $\rho_0 U^2 M$, which is smaller by a factor $\sqrt{M}$ when $M \ll 1$. The increased amplitude in two dimensions is a consequence of the infinite extent of the vortex source parallel to the cylinder. At any particular retarded time $t - |\mathbf{x}|/c_0$, the acoustic amplitude has the double-lobed directivity pattern illustrated in Fig. 1.7.1b for a dipole. Because $\Omega < 0$, the peaks of these lobes at a fixed distance $|\mathbf{x}|$ from the cylinder rotate in the clockwise direction at angular velocity $|\Omega|$ following the orbiting vortex, but with a phase lag of $\pi/4$ radians. The reader can confirm that the instantaneous force exerted on the fluid by the cylinder lies in the direction of the vector $\mathbf{x}_0(t)$

joining the center of the cylinder to the vortex. The radiation peak therefore also lags by $\pi/4$ the peak in the retarded surface force.

6.2.2 Example 2: Sound Produced by Vortex Motion near a Half-Plane (Crighton 1972)

The trajectory of a line vortex of strength Γ interacting with a rigid half-plane is shown in Fig. 4.6.2 for ideal motion at low Mach number. The sound generated by the vortex is calculated using the two-dimensional compact Green's function (3.9.9), which can be written

$$G_1(\mathbf{x}, \mathbf{y}, t - \tau) \approx \frac{\sin(\theta/2)\varphi^*(\mathbf{y})}{\pi\sqrt{|\mathbf{x}|}} \delta\left(t - \tau - \frac{|\mathbf{x}|}{c_0}\right), \quad |\mathbf{x}| \to \infty,$$

$$\varphi^*(\mathbf{y}) = \sqrt{r_0}\sin(\theta_0/2), \quad (6.2.7)$$

where $\mathbf{x} = |\mathbf{x}|(\cos\theta, \sin\theta)$, the coordinates being defined as in Fig. 4.6.2. This is applicable when the distance r_0 from the edge of the source at $(y_1, y_2) = r_0(\cos\theta_0, \sin\theta_0)$ is much smaller than the acoustic wavelength. For a line vortex at (r_0, θ_0) the characteristic frequency $\sim v/r_0$, where v is the vortex translational velocity. The wavelength is therefore of order

$$\frac{r_0}{v} \times c_0 = \frac{r_0}{M} \gg r_0 \quad \text{for } M = \frac{v}{c_0} \ll 1,$$

so that low Mach number motion is sufficient to ensure that the wavelength of the sound is much larger than the vortex distance from the edge.

Thus, adopting the notation of (6.2.3) and applying (6.1.1) in two dimensions (Green's function being given by (6.2.7)), we find

$$p(\mathbf{x}, t) \approx \frac{-\rho_0\Gamma\sin(\theta/2)}{\pi\sqrt{|\mathbf{x}|}} \int \mathbf{k} \wedge \frac{d\mathbf{x}_0}{d\tau}(\tau) \cdot \nabla\varphi^*(\mathbf{y})\delta(\mathbf{y} - \mathbf{x}_0(\tau))$$

$$\times \delta\left(t - \tau - \frac{|\mathbf{x}|}{c_0}\right) dy_1 \, dy_2 \, d\tau$$

$$= \frac{-\rho_0\Gamma\sin(\theta/2)}{\pi\sqrt{|\mathbf{x}|}} \left[\mathbf{k} \wedge \frac{d\mathbf{x}_0}{dt} \cdot \nabla\varphi^*\right], \quad |\mathbf{x}| \to \infty, \quad (6.2.8)$$

where on the second line the term in the square brackets is evaluated at the retarded position $\mathbf{x}_0(t - |\mathbf{x}|/c_0)$ of the vortex.

Now φ^* is the velocity potential of an ideal flow around the edge of the half-plane in the anticlockwise sense (with streamlines as in Fig. 6.2.3). It is

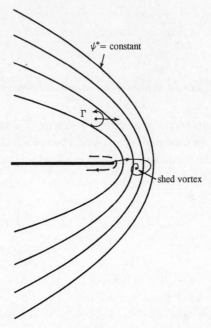

$\psi^* = $ constant

Γ

shed vortex

Fig. 6.2.3.

the real part of the complex potential

$$w = \varphi^* + i\psi^* = -i\sqrt{z}, \quad z = y_1 + iy_2,$$

where φ^* and the stream function ψ^* satisfy the *Cauchy–Riemann* equations (see Section 4.5)

$$\frac{\partial \varphi^*}{\partial y_1} = \frac{\partial \psi^*}{\partial y_2}, \quad \frac{\partial \varphi^*}{\partial y_2} = -\frac{\partial \psi^*}{\partial y_1}.$$

A simple calculation shows that

$$\mathbf{k} \wedge \nabla \varphi^* = \nabla \psi^*,$$

and therefore that

$$\mathbf{k} \wedge \frac{d\mathbf{x}_0}{dt} \cdot \nabla \varphi^* = -\frac{d\mathbf{x}_0}{dt} \wedge \mathbf{k} \cdot \nabla \varphi^* = -\frac{d\mathbf{x}_0}{dt} \cdot \mathbf{k} \wedge \nabla \varphi^* = -\frac{d\mathbf{x}_0}{dt} \cdot \nabla \psi^*.$$

The acoustic pressure can therefore be put in the form

$$p(\mathbf{x}, t) \approx \frac{\rho_0 \Gamma \sin(\theta/2)}{\pi \sqrt{|\mathbf{x}|}} \left[\frac{d\mathbf{x}_0}{dt} \cdot \nabla \psi^* \right] \equiv \frac{\rho_0 \Gamma \sin(\theta/2)}{\pi \sqrt{|\mathbf{x}|}} \left[\frac{D\psi^*}{Dt} \right], \quad |\mathbf{x}| \to \infty,$$

$$(6.2.9)$$

where $[D\psi^*/Dt]$ is evaluated at the retarded position of the vortex.

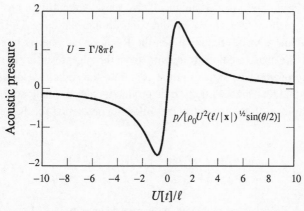

Fig. 6.2.4.

The stream function ψ^* is constant on each of the parabolic streamlines of the ideal flow around the edge defined by φ^* (Fig. 6.2.3). A vortex that translated along one of these streamlines would be silent; the actual edge-generated sound depends on the rate at which the trajectory of the vortex cuts across the streamlines of this ideal edge flow. This is found as follows

$$\psi^* = -\sqrt{r_0} \cos\left(\frac{\theta_0}{2}\right).$$

Therefore,

$$\frac{D\psi^*}{Dt} = -\frac{1}{2\sqrt{r_0}} \frac{dr_0}{dt} \cos\left(\frac{\theta_0}{2}\right) + \frac{\sqrt{r_0}}{2} \sin\left(\frac{\theta_0}{2}\right) \frac{d\theta_0}{dt},$$

where r_0, θ_0 are given by (4.6.7) (in which r is replaced by r_0 and θ by θ_0). Performing the calculations we find (see Fig. 6.2.4)

$$p(\mathbf{x}, t) \approx \frac{\rho_0 \Gamma^2}{(4\pi\ell)^2} \left(\frac{\ell}{|\mathbf{x}|}\right)^{\frac{1}{2}} \sin\left(\frac{\theta}{2}\right) \left[\frac{\Gamma t/8\pi\ell^2}{[1 + (\Gamma t/8\pi\ell^2)^2]^{5/4}}\right]_{t-|\mathbf{x}|/c_0}, \quad |\mathbf{x}| \to \infty,$$

where ℓ is the distance of closest approach of the vortex to the edge (where it crosses the x_1 axis at time $t = 0$ in Fig. 4.6.2).

6.3 Influence of Vortex Shedding

We can use Equation (6.2.9) to form a qualitative picture of the influence of vortex shedding on sound generation. Let the circulation of the vortex Γ in

Fig. 6.2.3 be in the indicated anticlockwise sense, so that fluid near the edge is induced to flow in a *clockwise* direction around the edge, as implied by the dashed curve in the figure. When the Reynolds number is large, inertial forces actually cause the flow to *separate* from the edge, resulting in the release of vorticity of *opposite* sign from the edge into the wake. Let us assume for simplicity that this shed vorticity rolls up into a concentrated core of strength Γ_s. Equation (6.2.9) then supplies the following prediction for the net acoustic pressure

$$p(\mathbf{x}, t) \approx \frac{\rho_0 \sin(\theta/2)}{\pi \sqrt{|\mathbf{x}|}} \left(\Gamma \left[\frac{D\psi^*}{Dt} \right]_\Gamma + \Gamma_s \left[\frac{D\psi^*}{Dt} \right]_{\Gamma_s} \right), \quad |\mathbf{x}| \to \infty,$$

where the derivatives are evaluated at the retarded positions of Γ and Γ_s respectively. Both vortices translate across the curves $\psi^* = $ constant in the direction of decreasing ψ^*, and the derivatives therefore have the same sign. Hence, because Γ and Γ_s have opposite signs, *sound produced by the shed vortex will tend to cancel the edge-generated sound attributable to the incident vortex Γ alone.*

This conclusion is applicable to a wide range of fluid–structure interactions. A typical interaction involves a bounded region of vorticity, called a 'gust,' swept along in a nominally steady mean flow. The localized velocity field of the gust is determined by the Biot–Savart formula (4.3.1). At high mean flow speeds it is sometimes permissible to neglect changes in the relative configuration of the vorticity distribution during its convection past an observation point, the vorticity is then said to be frozen (at least temporarily) and the induced velocity is steady in a frame translating with the gust. If the mean flow carries the gust past the surface S of a stationary body, the free field induced velocity determined by the Biot–Savart integral is said to produce an upwash velocity on S; the actual velocity consists of the upwash velocity augmented by the velocity field required to satisfy the no-slip condition on S. (In an ideal fluid only the normal component of the upwash velocity is cancelled on S.)

When a gust convects past a stationary airfoil the high Reynolds number surface force (responsible for the sound) is given by (see (4.4.4))

$$F_i = -\rho_0 \int_V \nabla Y_i(\mathbf{y}) \cdot (\boldsymbol{\omega} \wedge \mathbf{v})(\mathbf{y}, t) \, d^3\mathbf{y}, \quad Y_i = y_i - \varphi_i^*(\mathbf{y}). \quad (6.3.1)$$

The vector ∇Y_i is the velocity of an ideal flow past the airfoil that has unit speed in the i direction at large distances from the airfoil. It is singular (or very large) at the edges of the airfoil. These singularities have the following significance, when the vorticity length scale is small compared to the airfoil chord the principal contribution to the integral is from vorticity in the neighborhoods of the singularities. For example, for the strip airfoil of Fig. 3.6.3

$$Y_2 = \mathrm{Re}(-i\sqrt{z^2 - a^2}), \quad z = y_1 + iy_2,$$

and ∇Y_2 becomes infinite at the leading and trailing edges $z = \mp a$. An incident, small-scale gust convecting in the y_1 direction in a mean flow at speed U would in practice induce shedding from the trailing edge at $z = a$. When this shedding is ignored the force calculated from (6.3.1) has two principal components, respectively from gust elements near the leading and trailing edges. To calculate the overall force, however, it is necessary to include the contribution from the shed vorticity, which affects the motion only near the trailing edge when the length scale of the wake vorticity is small. In the linearized treatment of this case (discussed in more detail below), when both the gust and wake vorticity are taken to convect at the *same* mean velocity U, it is known from unsteady aerodynamics that the force component produced by the wake is equal and opposite to that generated by the gust at the trailing edge (Sears 1941).

The effect of this cancellation can be approximated without calculating any details of the shed vorticity. This is accomplished by formally deleting the trailing edge singularity from ∇Y_2, and then ignoring the contribution to the integral (6.3.1) from the shed vorticity. To understand this observe that, because the value of the integral is dominated by vorticity near the edges, it is only the behaviors of Y_2 near these edges that must be retained in the integrand, and Y_2 can therefore be replaced by the leading order terms in its expansions about the edges. For the strip airfoil of Fig. 3.6.3, we would write

$$Y_2 = \mathrm{Re}(-i\sqrt{z - a}\sqrt{z + a}) \sim \mathrm{Re}(\sqrt{2a}\sqrt{z + a}) + \mathrm{Re}(-i\sqrt{2a}\sqrt{z - a}). \tag{6.3.2}$$

The last term is singular at the trailing edge; it is deleted and the following approximation is used in (6.3.1) with the wake vorticity ignored:

$$Y_2 \sim \mathrm{Re}(\sqrt{2a}\sqrt{z + a}), \tag{6.3.3}$$

where the branch cut for $\sqrt{z + a}$ is taken along the real axis from $z = -a$ to $z = +\infty$.

6.3.1 Example: Surface Force Produced by a Periodic Gust

To illustrate the procedure consider incompressible flow parallel to the airfoil of Fig. 6.3.1 at speed U in the x_1 direction, in which a time harmonic vortex

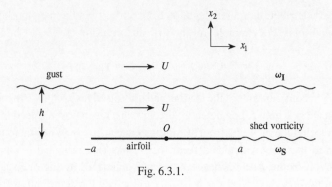

Fig. 6.3.1.

sheet of vorticity

$$\boldsymbol{\omega}_{\mathrm{I}} = \gamma \mathbf{k} \delta(x_2 - h) e^{-i\omega(t - x_1/U)}, \quad h > 0, \quad \omega > 0,$$

is convected past the airfoil at perpendicular distance h, where γ is the *circulation per unit length* of the sheet. The vortex sheet can be regarded as an elementary model of a periodic wake behind a small diameter circular cylinder upstream of the airfoil. The gust upwash velocity induces the shedding of vorticity ω_{S} from the trailing edge of the airfoil. When the *reduced frequency* $\omega a/U$ is large the hydrodynamic wavelength $2\pi U/\omega$ of the gust and wake is much smaller than the airfoil chord, and the surface force is produced primarily by the gust interaction with the leading edge at $x_1 = -a$.

The net force F_2 (per unit span) on the fluid in the x_2 direction can therefore be calculated from (6.3.1) by setting $\omega = \omega_{\mathrm{I}}$, where

$$\boldsymbol{\omega}_{\mathrm{I}} \wedge \mathbf{v} = \gamma U \mathbf{j} \delta(x_2 - h) e^{-i\omega(t - x_1/U)},$$

($\mathbf{j}$ being a unit vector in the x_2 direction) and by replacing Y_2 by the right-hand side of (6.3.3):

$$
\begin{aligned}
F_2 &= -\rho_0 \gamma U \int_{-\infty}^{\infty} \frac{\partial Y_2}{\partial y_2} (\mathbf{y}) \delta(y_2 - h) e^{-i\omega(t - y_1/U)} \, dy_1 \, dy_2 \\
&\approx -\rho_0 \gamma U \sqrt{2a} \int_{-\infty}^{\infty} \mathrm{Re} \left(\frac{i}{2\sqrt{y_1 + ih + a}} \right) e^{-i\omega(t - y_1/U)} \, dy_1 \\
&= -\rho_0 \gamma U a \left(\frac{\pi U}{2i\omega a} \right)^{\frac{1}{2}} e^{-\omega h/U - i\omega(t + a/U)}.
\end{aligned}
\tag{6.3.4}
$$

The force can also be calculated exactly from linearized thin airfoil theory with proper account taken of vortex shedding. This is the gust loading problem of classical aerodynamics (Sears 1941). The linear theory wake is treated as a vortex sheet downstream of the edge, whose elements convect at the mean stream velocity U. The strength of the vortex sheet is determined by imposing the *Kutta condition* that the pressure (and velocity) should be finite at the edge (Crighton 1985). For arbitrary values of the reduced frequency $\omega a / U$ it is found that

$$F_2 = \pi i \rho_0 \gamma U a \mathcal{S} \left(\frac{\omega a}{U} \right) e^{-\omega h / U - i \omega t}, \qquad (6.3.5)$$

where $\mathcal{S}(x)$ is the *Sears function*, which is defined in terms of the Hankel functions $H_0^{(1)}$ and $H_1^{(1)}$ by

$$\mathcal{S}(x) = \frac{2}{\pi x \left[H_0^{(1)}(x) + i H_1^{(1)}(x) \right]}. \qquad (6.3.6)$$

In the limit of high reduced frequency

$$\mathcal{S} \left(\frac{\omega a}{U} \right) \sim \left(\frac{i U}{2 \pi \omega a} \right)^{\frac{1}{2}} e^{-i \omega a / U},$$

and the substitution of this into (6.3.5) yields the prediction (6.3.4) determined by the leading edge singularity of Y_2. The plots in Fig. 6.3.2 of the real and imaginary parts of $\mathcal{S}(\omega a / U)$ and its asymptotic limit show that the approximation (6.3.4) and the exact value (6.3.5) of the surface force agree when $\omega a / U > 1$.

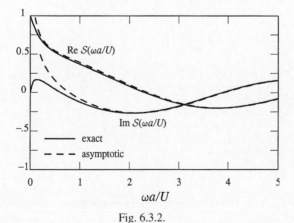

Fig. 6.3.2.

Linear theory does not permit the corresponding force component at the leading edge to be removed by vorticity production at the edge, because it requires vorticity shed there to be swept over the rigid surface of the airfoil on which it cannot influence the force because $\omega \wedge \mathbf{v} \cdot \nabla Y_2 \equiv 0$.

6.4 Blade–Vortex Interaction Noise in Two Dimensions

A rigid, two-dimensional airfoil of chord $2a$ occupies $-a < x_1 < a$, $x_2 = 0$ in the presence of a uniform mean stream at speed U in the positive x_1 direction. A line vortex of strength Γ parallel to the airfoil span convects at the mean flow velocity at a constant distance h above the airfoil (Fig. 6.4.1). This is the approximation of linearized thin airfoil theory, which is applicable when

$$\frac{\Gamma}{h} \ll U,$$

that is, when the influence on the vortex trajectory of the induced velocity ($\sim \Gamma / h$) of image vortices in the body of the airfoil is negligible. As the vortex passes the airfoil new vorticity is shed from the trailing edge into the airfoil wake, which is assumed to consist of a vortex sheet stretching along the x_1 axis from $x_1 = a$ to $x_1 = +\infty$.

Let us first consider the potential flow interaction of the vortex and airfoil, when no account is taken of vortex shedding. Suppose the vortex passes above the midchord of the airfoil at time $t = 0$, then

$$\omega = \Gamma \mathbf{k} \delta(x_1 - Ut)\delta(x_2 - h), \qquad \mathbf{v} = U\mathbf{i}.$$

Hence,

$$\omega \wedge \mathbf{v} = \Gamma U \mathbf{j} \delta(x_1 - Ut)\delta(x_2 - h), \tag{6.4.1}$$

where $\mathbf{i}$ and $\mathbf{j}$ are unit vectors in the x_1 and x_2 directions. The acoustic pressure

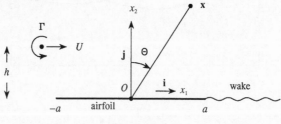

Fig. 6.4.1.

generated when the wake is ignored is given by (6.2.1) with

$$Y_1 = y_1, \qquad Y_2 = \text{Re}\,(-i\sqrt{z^2 - a^2}), \quad z = y_1 + iy_2.$$

Thus, $\nabla Y_1 = \mathbf{i}$ and $\boldsymbol{\omega} \wedge \mathbf{v} \cdot \nabla Y_1 \equiv 0$, and (6.2.1) reduces to

$$
\begin{aligned}
p(\mathbf{x}, t) &\approx \frac{-\rho_0 \Gamma U x_2}{2\pi \sqrt{2c_0}|\mathbf{x}|^{\frac{3}{2}}} \frac{\partial}{\partial t} \int_{-\infty}^{t-|\mathbf{x}|/c_0} \frac{\partial Y_2}{\partial y_2}(U\tau, h) \frac{d\tau}{\sqrt{t - \tau - |\mathbf{x}|/c_0}}, \\
&= \frac{-\rho_0 \Gamma U \cos \Theta}{2\pi \sqrt{2c_0}|\mathbf{x}|^{\frac{1}{2}}} \frac{\partial}{\partial t} \int_{-\infty}^{t-|\mathbf{x}|/c_0} \frac{\partial Y_2}{\partial y_2}(U\tau, h) \frac{d\tau}{\sqrt{t - \tau - |\mathbf{x}|/c_0}},
\end{aligned}
$$
$$|\mathbf{x}| \to \infty, \quad (6.4.2)$$

where $\Theta = \cos^{-1}(x_2/|\mathbf{x}|)$ is the angle between the radiation direction $\mathbf{x}$ and the normal to the airfoil (the x_2 axis). The sound can be attributed to a dipole source orientated in the x_2 direction.

The radiation produced when a vortex passes very close to the airfoil (so that $h \ll a$) is likely to be particularly intense. The dominant interactions occur as the vortex passes the edges, where the time scales of the motions $\sim h/U$. Thus, the characteristic frequency

$$\omega \sim \frac{U}{h} \quad \text{and the reduced frequency} \quad \frac{\omega a}{U} \sim \frac{a}{h} \gg 1.$$

We may therefore regard the leading and trailing edges as independent sources of sound, and calculate their individual contributions by using the local approximation (6.3.2). For the acoustic pressure p_{LE}, say, produced at the leading edge $(x_1 = -a)$ we take

$$Y_2 \sim \text{Re}\,(\sqrt{2a}\sqrt{z + a}),$$

so that (6.4.2) becomes

$$
\begin{aligned}
p_{\text{LE}} &\approx \frac{-\rho_0 \Gamma U \cos \Theta}{2\pi \sqrt{2c_0}|\mathbf{x}|^{\frac{1}{2}}} \frac{\partial}{\partial t} \int_{-\infty}^{t-|\mathbf{x}|/c_0} \text{Re}\left(\frac{i\sqrt{2a}}{2(U\tau + ih + a)^{\frac{1}{2}}} \right) \\
&\quad \times \frac{d\tau}{\sqrt{t - \tau - |\mathbf{x}|/c_0}}, \quad |\mathbf{x}| \to \infty.
\end{aligned}
$$

To evaluate the integral, make the substitution $\mu = \sqrt{t - \tau - |\mathbf{x}|/c_0}$ and perform the differentiation with respect to time. Then,

$$p_{LE} \approx \frac{-\rho_0 \Gamma U \cos \Theta}{2\pi \sqrt{c_0}} \left(\frac{a}{|\mathbf{x}|} \right)^{\frac{1}{2}} \frac{\partial}{\partial t} \int_0^\infty \text{Re} \left[\frac{i}{(U[t] + ih + a - U\mu^2)^{\frac{1}{2}}} \right] d\mu$$

$$= \frac{\rho_0 \Gamma U^2 \cos \Theta}{4\pi \sqrt{c_0}} \left(\frac{a}{|\mathbf{x}|} \right)^{\frac{1}{2}} \int_0^\infty \text{Re} \left[\frac{i}{(U[t] + ih + a - U\mu^2)^{\frac{3}{2}}} \right] d\mu,$$

$$\text{where} \quad [t] = t - \frac{|\mathbf{x}|}{c_0}.$$

The additional substitution $\mu = 1/\xi$ transforms the integrand into an exact differential, leading finally to

$$p_{LE} \approx \frac{\rho_0 \Gamma U^2 \cos \Theta}{4\pi \sqrt{c_0}} \left(\frac{a}{|\mathbf{x}|} \right)^{\frac{1}{2}} \int_0^\infty \text{Re} \left[\frac{i\xi}{[(U[t] + ih + a)\xi^2 - U]^{\frac{3}{2}}} \right] d\xi$$

$$= \frac{\rho_0 \Gamma U \sqrt{M} \cos \Theta}{4\pi a} \left(\frac{a}{|\mathbf{x}|} \right)^{\frac{1}{2}} \frac{\left(\frac{U[t]}{a} + 1 \right)}{\left(\frac{U[t]}{a} + 1 \right)^2 + \left(\frac{h}{a} \right)^2}, \quad |\mathbf{x}| \to \infty, \quad (6.4.3)$$

where $M = U/c_0$.

The corresponding nondimensional pressure signature

$$p_{LE} \left/ \frac{\rho_0 \Gamma U \sqrt{M} \cos \Theta}{4\pi a} \left(\frac{a}{|\mathbf{x}|} \right)^{\frac{1}{2}} \right.$$

is plotted in Fig. 6.4.2 as the solid curve when $h/a = 0.2$. The pressure field

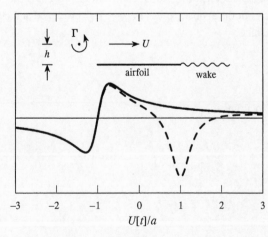

Fig. 6.4.2.

is generated predominantly as the vortex passes the leading edge of the airfoil at the retarded time $[t] = -a/U$, with characteristic frequency $\omega \sim U/h$. According to Section 6.3, at high reduced frequencies the sound pressure p_{TE}, say, generated by the potential flow interaction of the vortex with the trailing edge is cancelled by that produced by the wake vorticity; the solid curve in Fig. 6.4.2 is therefore representative of the whole of the radiation produced by the blade–vortex interaction.

To determine p_{TE} the calculation described above for p_{LE} is repeated after setting

$$Y_2 = \mathrm{Re}(-i\sqrt{2a}\sqrt{z-a})$$

in (6.4.2), leading to

$$
\begin{aligned}
p_{TE} &\approx \frac{-\rho_0 \Gamma U \cos \Theta}{4\pi \sqrt{c_0}} \left(\frac{a}{|\mathbf{x}|} \right)^{\frac{1}{2}} \frac{\partial}{\partial t} \int_{-\infty}^{t-|\mathbf{x}|/c_0} \mathrm{Re}\left(\frac{1}{(U\tau + ih - a)^{\frac{1}{2}}} \right) \\
&\quad \times \frac{d\tau}{\sqrt{t - \tau - |\mathbf{x}|/c_0}} \\
&= \frac{-\rho_0 \Gamma U \sqrt{M} \cos \Theta}{4\pi a} \left(\frac{a}{|\mathbf{x}|} \right)^{\frac{1}{2}} \frac{\left(\frac{h}{a} \right)}{\left(\frac{U[t]}{a} - 1 \right)^2 + \left(\frac{h}{a} \right)^2}, \quad |\mathbf{x}| \to \infty. \quad (6.4.4)
\end{aligned}
$$

This is large at the retarded times during which the vortex is close to the trailing edge. Thus, when the contribution from the wake is ignored (which is equal and opposite to p_{TE}) the overall acoustic pressure signature is given nondimensionally by

$$
(p_{LE} + p_{TE}) \bigg/ \frac{\rho_0 \Gamma U \sqrt{M} \cos \Theta}{4\pi a} \left(\frac{a}{|\mathbf{x}|} \right)^{\frac{1}{2}},
$$

which is plotted as the broken line curve in Fig. 6.4.2.

The leading and trailing edge generated components p_{LE} and p_{TE} have different waveforms, even though they are produced by the vortex interacting with geometrically identical airfoil edges. This is because the integral in (6.4.2) determines the acoustic pressure at the retarded time $[t]$ in terms of interactions between the vortex and the airfoil *at all earlier retarded times*; it is a further consequence of the two-dimensional character of the acoustic sources, according to which, after the first arrival of sound from the nearest point on the source, additional contributions to the sound continue to be received *indefinitely in time* from more distant parts of the source.

Problems 6

1. Starting from the formula

$$
G \approx \frac{\bar{\mathbf{x}} \cdot \mathbf{Y}}{2\pi c_0} \frac{\partial}{\partial t} \int_0^\infty \delta\left(t - \tau - \frac{\sqrt{|\bar{\mathbf{x}}|^2 + \xi^2}}{c_0}\right) \frac{d\xi}{(|\bar{\mathbf{x}}|^2 + \xi^2)},
$$

derive the far-field approximation (6.1.5) for the dipole component of the two-dimensional compact Green's function by writing

$$
\delta\left(t - \tau - \frac{\sqrt{|\bar{\mathbf{x}}|^2 + \xi^2}}{c_0}\right) = \frac{1}{2\pi} \int_{-\infty}^\infty e^{-i\omega\left(t - \tau - \frac{\sqrt{|\bar{\mathbf{x}}|^2 + \xi^2}}{c_0}\right)} d\omega,
$$

and applying Formula (5.2.11).

2. Investigate the production of sound by the low Mach number motion of a line vortex of strength Γ that is parallel to a rigid circular cylinder of radius a whose axis coincides with the x_3 coordinate axis (c.f., Section 4.6, Example 2). Assume that there is no net circulation around the cylinder, and that there is a mean flow past the cylinder which has speed U in the x_1 direction when $|x_1| \gg a$. If the vortex is initially far upstream of the cylinder at a distance h from the x_1 axis, examine the production of sound for different values of the nondimensional parameter Γ/Uh.

3. A line vortex of strength Γ is parallel to a rigid airfoil occupying $-a < x_1 < a$, $x_2 = 0$, $-\infty < x_3 < \infty$, in the presence of a mean flow at speed U in the x_1 direction. The vortex is initially far upstream of the airfoil at a vertical stand-off distance h above the plane of the airfoil. There is no net circulation around the airfoil. If the motion occurs at very small Mach number, calculate the sound produced when the vortex passes the airfoil for different values of the nondimensional velocity ratio Γ/Uh. When $U \gg \Gamma/h$, estimate the influence on the sound of vortex shedding from the trailing edge of the airfoil.

4. A line vortex of strength Γ is parallel to a rigid airfoil occupying

$$
-a < x_1 < a, \qquad x_2 = 0, \qquad -\infty < x_3 < \infty
$$

in fluid at rest at infinity. The vortex is in periodic motion around the airfoil under the influence of image vortices in the absence of a mean circulation around the airfoil. Calculate the sound produced when the motion occurs at a very small Mach number, and show that it can be attributed to two dipole sources orientated in the x_1 and x_2 directions. Explain the significance of these sources in terms of the corresponding components of the unsteady force between the fluid and airfoil.

5. A line vortex of strength Γ traverses a path of the kind illustrated in the figure past a two-dimensional, thin rigid barrier of length d at right angles to a plane wall at $x_2 = 0$. There is a low Mach number mean potential flow over the barrier that has speed U parallel to the wall at large distances from the barrier. If the distance of the vortex from the wall is h when the vortex is far upstream of the barrier, calculate the sound produced as the vortex passes the barrier for different values of Γ/Uh. Explain what happens when $\Gamma/h \ll U$. Discuss the forces exerted on the barrier by the flow, and how they contribute to the radiation.

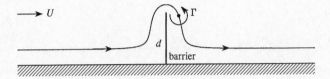

7

Problems in Three Dimensions

7.1 Linear Theory of Vortex–Airfoil Interaction Noise

Consider an inhomogeneous field of vorticity, a gust, convected in high Reynolds number, homentropic flow past a stationary rigid airfoil (Fig. 7.1.1). The undisturbed flow has speed U in the x_1 direction, where the origin is at a convenient point within the airfoil, with x_3 along the span and x_2 vertically upward. The Mach number $M = U/c_0$ is sufficiently small that convection of sound by the flow can be ignored, and the airfoil chord can be assumed to be acoustically compact.

The vortex sound source $\text{div}(\boldsymbol{\omega} \wedge \mathbf{v})$ includes vorticity in the gust together with any vorticity shed from the airfoil, either in response to excitation by the gust, or as tip vortices responsible for the mean lift. The problem can be linearized by assuming that $u \ll U$, where $\text{curl } \mathbf{u} = \boldsymbol{\omega}$, that is, by requiring the gust-induced velocity, and the perturbation velocities caused by airfoil thickness, twist, camber, and the angle of attack, to be small. When $\text{div}(\boldsymbol{\omega} \wedge \mathbf{v})$ is expanded about the undisturbed mean flow, only the gust vorticity and additional vorticity shed when the gust encounters the airfoil contribute to the acoustic radiation to first order. In other words, thickness, twist, camber, and angle of attack may ignored, and the airfoil regarded as a rigid lamina in the plane $x_2 = 0$. In this approximation, quadrupoles are neglected and vorticity convects as a *frozen* pattern of vortex filaments at the undisturbed mean stream velocity $\mathbf{U} = (U, 0, 0)$. In particular, the wake vorticity is confined to a vortex sheet downstream of the trailing edge.

When convection of sound by the flow is neglected, the linearized form of the vortex sound equation (5.2.5) becomes

$$\left(\frac{1}{c_0^2} \frac{\partial^2}{\partial t^2} - \nabla^2 \right) B = \text{div}(\boldsymbol{\omega} \wedge \mathbf{U}), \qquad (7.1.1)$$

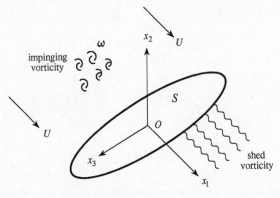

Fig. 7.1.1.

with solution

$$B(\mathbf{x}, t) = \frac{p(\mathbf{x}, t)}{\rho_0} \approx - \int (\boldsymbol{\omega} \wedge \mathbf{U})(\mathbf{y}, \tau) \cdot \frac{\partial G}{\partial \mathbf{y}}(\mathbf{x}, \mathbf{y}, t - \tau)\, d^3\mathbf{y}\, d\tau, \quad |\mathbf{x}| \to \infty,$$

(7.1.2)

where $\partial G/\partial y_2 = 0$ on both sides ($y_2 = \pm 0$) of the projection of the airfoil plan-form onto the y_1, y_3 plane. At sufficiently small Mach numbers G may be approximated by Green's function for an airfoil of compact chord.

The sound produced when a localized, high Reynolds number *frozen* gust $\boldsymbol{\omega}(\mathbf{x} - \mathbf{U}t) \equiv \boldsymbol{\omega}(x_1 - Ut, x_2, x_3)$ is swept past the airfoil of Fig. 7.1.1 is therefore given by (5.4.4) with the convection velocity $\mathbf{v}$ replaced by $\mathbf{U}$, and the Kirchhoff vector $\mathbf{Y}$ by

$$Y_1 = y_1, \qquad Y_2 = y_2 - \varphi_2^*(\mathbf{y}), \qquad Y_3 = y_3.$$

Thus, because $\int \boldsymbol{\omega}\, d^3\mathbf{y} \equiv 0$, Equation (5.4.4) reduces to

$$p(\mathbf{x}, t) \approx \frac{-\rho_0 U \cos \Theta}{4\pi c_0 |\mathbf{x}|} \frac{\partial}{\partial t} \int \left[\omega_3 \frac{\partial Y_2}{\partial y_2} - \omega_2 \frac{\partial Y_2}{\partial y_3} \right]_{t - |\mathbf{x}|/c_0} d^3\mathbf{y}, \quad |\mathbf{x}| \to \infty,$$

(7.1.3)

where $\Theta = \cos^{-1}(x_2/|\mathbf{x}|)$ is the angle between the radiation direction and the normal to the airfoil, and the origin is taken in the airfoil within the interaction region.

Let the interaction occur at an inboard location where the chord may be regarded as constant, with both the leading and trailing edges at right angles to the mean flow (so that $\partial Y_2/\partial y_3 \ll \partial Y_2/\partial y_2$). The planform in the interaction

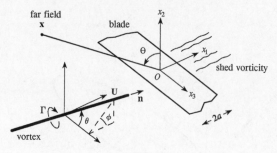

Fig. 7.2.1.

region is then locally the same as that of the two-dimensional airfoil of Fig. 7.2.1, and Y_2 can be approximated as in (3.9.2) for constant $a \equiv a(y_3)$, such that $2a$ is equal to the local chord of the airfoil. Then, (7.1.3) becomes

$$p(\mathbf{x}, t) \approx \frac{-\rho_0 U \cos \Theta}{4\pi c_0 |\mathbf{x}|} \frac{\partial}{\partial t} \int \left[\omega_3 \frac{\partial Y_2}{\partial y_2} \right]_{t - |\mathbf{x}|/c_0} d^3\mathbf{y}, \quad |\mathbf{x}| \to \infty, \quad (7.1.4)$$

which reveals that only the spanwise component of vorticity contributes to the production of sound.

According to Equation (5.4.3) (in which $dU_j/dt = 0$ for a stationary airfoil), this result can also be expressed in the form

$$p(\mathbf{x}, t) \approx \frac{\cos \Theta}{4\pi c_0 |\mathbf{x}|} \frac{\partial F_2}{\partial t} \left(t - \frac{|\mathbf{x}|}{c_0} \right), \quad |\mathbf{x}| \to \infty,$$

$$F_2(t) = -\rho_0 U \int \omega_3(\mathbf{y}, t) \frac{\partial Y_2}{\partial y_2}(\mathbf{y}) \, d^3\mathbf{y}, \quad (7.2.5)$$

where $-F_2$ is the unsteady airfoil *lift* during the interaction when the motion is regarded as incompressible, and the vorticity ω_3 includes contributions from the impinging gust together with any shed into the vortex sheet wake.

7.2 Blade–Vortex Interactions in Three Dimensions

The calculations can be performed explicitly for a gust in the form of a rectilinear *line vortex*. Let the vortex have circulation Γ and be orientated with its axis in the direction of the unit vector $\mathbf{n}$, as indicated in Fig. 7.2.1. The mean flow speed is sufficiently large that the vortex can be assumed to maintain its rectilinear form after being cut by the leading edge of the airfoil. Choose the origin on the airfoil midchord such that the axis of the vortex passes through the origin at time $t = 0$. Any point $\mathbf{x}$ on the vortex can then be represented in the parametric form

$$\mathbf{x} = (Ut, 0, 0) + s\mathbf{n}, \qquad -\infty < s < \infty, \qquad (7.2.1)$$

where s is distance measured along the vortex from its point of intersection with the plane of the airfoil. Then, if $\mathbf{s}_\perp$ denotes vector distance measured in the normal direction from the vortex axis,

$$\boldsymbol{\omega} = \Gamma \mathbf{n} \delta(\mathbf{s}_\perp),$$

where the polar angles θ, ϕ in Figure 7.2.1 define the orientation of the unit vector

$$\mathbf{n} = (\sin\theta\cos\phi, \sin\theta\sin\phi, \cos\theta), \quad 0 < \theta < \pi, \quad 0 < \phi < 2\pi.$$

Because of vortex shedding from the trailing edge, most of the sound is generated when the vortex is cut by the leading edge. The influence of the shed vorticity can be formally included by the procedure described in Sections 6.3 and 6.4 by expanding Y_2 about its singularity at the leading edge, that is, by setting

$$Y_2 \sim \text{Re}(\sqrt{2a}\sqrt{z+a}), \quad z = y_1 + iy_2,$$

with the branch cut for the square root taken along the z axis from $z = -a$ to $z = +\infty$. Thus, by recalling the Relation (7.2.1), Equation (7.1.4) becomes

$$
\begin{aligned}
p(\mathbf{x}, t) &\approx \frac{-n_3\rho_0\Gamma U \cos\Theta}{4\pi c_0|\mathbf{x}|} \frac{\partial}{\partial t} \text{Re} \int \left[\delta(\mathbf{s}_\perp) \left(\frac{i\sqrt{2a}}{2\sqrt{y_1 + iy_2 + a}} \right) \right]_{t - |\mathbf{x}|/c_0} d^2\mathbf{s}_\perp \, ds \\
&= \frac{-n_3\rho_0\Gamma U \cos\Theta}{4\sqrt{2}\pi c_0|\mathbf{x}|} \frac{\partial}{\partial t} \text{Re} \int_{-\infty}^{\infty} \left(\frac{i\sqrt{a}}{\sqrt{U[t] + s(n_1 + in_2) + a}} \right) ds \\
&= \frac{n_3\rho_0\Gamma U^2 \cos\Theta}{8\sqrt{2}\pi c_0|\mathbf{x}|} \text{Re} \int_{-\infty}^{\infty} \left(\frac{i\sqrt{a}}{(U[t] + s(n_1 + in_2) + a)^{\frac{3}{2}}} \right) ds \\
&= \frac{-n_3\rho_0\Gamma U^2 \cos\Theta}{4\sqrt{2}\pi c_0|\mathbf{x}|} \text{Re} \left[\frac{i\sqrt{a}}{(U[t] + s(n_1 + in_2) + a)^{\frac{1}{2}}(n_1 + in_2)} \right]_{-\infty}^{\infty},
\end{aligned}
$$

$$(7.2.2)$$

where $[t] = t - |\mathbf{x}|/c_0$ is the retarded time.

By referring to Fig. 7.2.2 it will be seen that the last line of (7.2.2) is zero when

$$U[t] + a < 0,$$

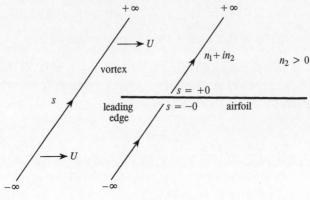

Fig. 7.2.2.

before the vortex is cut by the leading edge of the airfoil. At later times the integration along the vortex axis over the infinite range $-\infty < s < \infty$ must be split, as indicated in the figure, into the two parts $-\infty < s < -0, +0 < s < \infty$, because the square root in the last line of (7.2.2) is discontinuous across the airfoil; for example, when $n_2 > 0$ the square root is real and positive on the upper surface ($s = +0$) and real and negative at $s = -0$. Hence, the acoustic pressure becomes

$$p(\mathbf{x}, t) \approx \frac{\rho_0 \Gamma U M \cos \Theta}{2^{\frac{3}{2}} \pi |\mathbf{x}|} \frac{|\sin\phi|}{\tan \theta} \frac{H\left(\frac{U[t]}{a} + 1\right)}{\sqrt{\frac{U[t]}{a} + 1}}. \qquad (7.2.3)$$

The pressure pulse accordingly begins with a singular peak at the instant at which the vortex is severed by the leading edge of the airfoil at the retarded time $U[t]/a = -1$. The waveform is illustrated by the dotted curve in Fig. 7.2.3, which is a plot of

$$\frac{p(\mathbf{x}, t)}{(\rho_0 \Gamma U M \cos \Theta / |\mathbf{x}|)} = \frac{|\sin\phi|}{2^{\frac{3}{2}} \pi \tan \theta} \frac{H\left(\frac{U[t]}{a} + 1\right)}{\sqrt{\frac{U[t]}{a} + 1}} \quad \text{for } \theta = 85°, \phi = 90°.$$

The infinite singularity in the pressure is absent for a vortex of nonzero core radius R, say. If, for example, the vorticity is assumed to be distributed according to the Gaussian formula

$$\omega(\mathbf{x}) = \frac{\Gamma \mathbf{n} e^{-(s_\perp/R)^2}}{\pi R^2},$$

as a function of distance $s_\perp$ from the vortex axis, the acoustic pressure is found

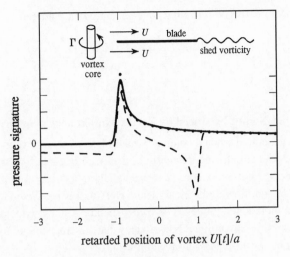

Fig. 7.2.3.

to be given by (Howe 1998a)

$$\frac{p(\mathbf{x}, t)}{(\rho_0 \Gamma U M \cos \Theta / |\mathbf{x}|)} \approx \frac{|\sin \phi|^{\frac{3}{2}}}{8 \tan \theta} \left(\frac{a}{\pi R} \right)^{\frac{1}{2}} \Im(\alpha), \quad |\mathbf{x}| \to \infty, \quad \text{for } R \ll a,$$
(7.2.4)

where

$$\Im(\alpha) = |\alpha|^{\frac{1}{2}} \left\{ I_{-\frac{1}{4}} \left(\frac{\alpha^2}{8} \right) + \text{sgn}(\alpha) I_{\frac{1}{4}} \left(\frac{\alpha^2}{8} \right) \right\} e^{-\alpha^2/8}, \quad \alpha = \frac{2a|\sin \phi|}{R \left(\frac{U[t]}{a} + 1 \right)},$$

and $I_{\pm \frac{1}{4}}$ are modified Bessel functions of the first kind. The pressure signature predicted by (7.2.4) for

$$R = 0.1a, \qquad \theta = 85°, \qquad \phi = 90°$$

is plotted as the solid curve in Fig. 7.2.3, and differs negligibly from the line vortex prediction when $U[t]/a > -1$.

The broken-line curve in Fig. 7.2.3 represents the pressure signature produced by the *potential flow* interaction between the finite core vortex and airfoil (i.e., when vortex shedding is ignored). It is an odd function of the retarded time $[t]$. The large negative peak produced as the vortex crosses the trailing edge (at $U[t]/a = 1$) is cancelled by an equal and opposite contribution generated by the wake. The reader can easily show that, when the finite size of the vortex core is ignored, the potential flow, trailing edge generated pressure pulse is given by

$$p(\mathbf{x}, t) \approx -\frac{\rho_0 \Gamma U M \cos \Theta}{2^{\frac{3}{2}} \pi |\mathbf{x}|} \frac{|\sin\phi|}{\tan\theta} \frac{H\left(1 - \frac{U[t]}{a}\right)}{\sqrt{1 - \frac{U[t]}{a}}}.$$

7.3 Sound Produced by Vortex Motion near a Sphere

The sound generated when a nominally rectilinear vortex is swept past a *compact* rigid body can also be treated in a linearized fashion, by assuming that each element of the vortex core is convected along a streamline of the steady undisturbed mean flow at the local mean velocity. It is not generally possible, however, to include the influence of vortex shedding in a satisfactory manner, except perhaps for streamlined body shapes that are amenable to treatment by the strip theory of unsteady aerodynamics.

To illustrate the procedure, consider a rigid sphere of radius a with center at the coordinate origin in the presence of a low Mach number *irrotational* mean flow which is in the x_1 direction at speed U for $|\mathbf{x}| \gg a$. The mean velocity at $\mathbf{x}$ can therefore be written

$$\mathbf{U} = U\nabla X_1(\mathbf{x}), \tag{7.3.1}$$

where $X_1(\mathbf{x})$ is the x_1 component of the Kirchhoff vector for the sphere, which has the general form (Table 3.9.1)

$$X_i = x_i \left(1 + \frac{a^3}{2|\mathbf{x}|^3}\right). \tag{7.3.2}$$

Suppose a line vortex of strength Γ is initially far upstream of the sphere and parallel to the x_3 axis at a distance h above the plane $x_2 = 0$. The vortex is convected toward the sphere by the mean flow. The part of the vortex that passes close to the sphere must evidently be deformed to pass around the sphere; more distant parts of the vortex (at $|x_3| \gg a$) are unaffected and remain parallel to the x_3 direction during the whole of the interaction. The shape of the distorted vortex will be symmetric with respect to the mid-plane $x_3 = 0$; the vortex element initially on $x_3 = 0$ remains on this plane of symmetry as it convects past the sphere along a mean streamline, as illustrated in Fig. 7.3.1, which shows the motion in the plane $x_3 = 0$.

The shape of the vortex at time t is determined by the solution of the equations

$$\frac{dx_1}{dt} = U\frac{\partial X_1}{\partial x_1}(\mathbf{x}), \qquad \frac{dx_2}{dt} = U\frac{\partial X_1}{\partial x_2}(\mathbf{x}), \qquad \frac{dx_3}{dt} = U\frac{\partial X_1}{\partial x_3}(\mathbf{x}),$$

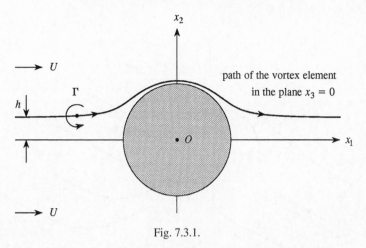

Fig. 7.3.1.

for each element of the vortex. If the undistorted parts of the vortex (at $|x_3| \gg a$) are assumed to convect across the plane $x_1 = 0$ at time $t = 0$, these equations are to be integrated subject to the initial conditions

$$x_1 = Ut, \qquad x_2 = h, \qquad x_3 = x_3^0 \qquad t \to -\infty,$$

where x_3^0 is the initial spanwise location of the vortex element.

In terms of the nondimensional variables

$$T = \frac{Ut}{a}, \qquad \bar{\mathbf{x}} = \frac{\mathbf{x}}{a},$$

the equations of motion of a point on the vortex are found to be

$$\frac{d\bar{x}_1}{dT} = 1 + \frac{\bar{x}_2^2 + \bar{x}_3^2 - 2\bar{x}_1^2}{2(\bar{x}_1^2 + \bar{x}_2^2 + \bar{x}_3^2)^{\frac{5}{2}}}, \qquad \frac{d\bar{x}_2}{dT} = \frac{-3\bar{x}_1\bar{x}_2}{2(\bar{x}_1^2 + \bar{x}_2^2 + \bar{x}_3^2)^{\frac{5}{2}}},$$

$$\frac{d\bar{x}_3}{dT} = \frac{-3\bar{x}_1\bar{x}_3}{2(\bar{x}_1^2 + \bar{x}_2^2 + \bar{x}_3^2)^{\frac{5}{2}}}.$$

These are solved (for example, by the Runge–Kutta method described in Section 4.6) by starting the integration at $T = -10$, say. It can be safely assumed that the sphere has no perceptible influence on vortex elements initially located at $|\bar{x}_3| > 10$. Because the motion is symmetric about $\bar{x}_3 = 0$ the solutions are required only for the $N + 1$ vortex elements with respective the initial positions

$$\bar{x}_1 = T, \qquad \bar{x}_2 = \frac{h}{a}, \qquad \bar{x}_3 = \bar{x}_3^n \quad \text{at } T = -10,$$

where $\bar{x}_3^n = 10n/N$, $0 \le n \le N$, and N is a suitably large integer.

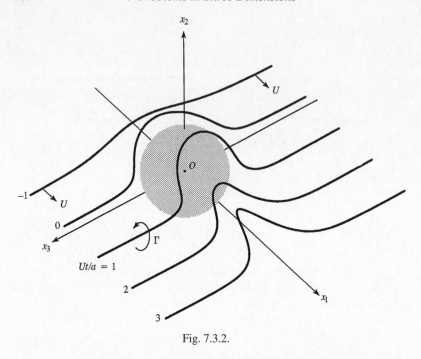

Fig. 7.3.2.

Figure 7.3.2 illustrates successive calculated positions of the vortex with increasing values of the time $T = Ut/a$ for the case $h/a = 0.2$. The distortion of the vortex first becomes evident at about $T = -2$. The hairpin loop is formed because the translation velocities of vortex elements close to the sphere are small in the neighborhood of the stagnation points just in front and just to the rear of the sphere; the accelerated motion over the upper surface of the sphere is insufficient to counteract the formation of the loop. In reality, of course, the motion would be strongly influenced by large self-induced velocities, by image vorticity in the sphere, and by viscous diffusion of vorticity from the vortex and from the surface of the sphere, none of which is accounted for in the present calculation.

The sound generated during this potential flow interaction can be calculated using the Formula (5.4.4) when the sphere is compact:

$$p(\mathbf{x}, t) \approx \frac{-\rho_0 x_j}{4\pi c_0 |\mathbf{x}|^2} \frac{\partial}{\partial t} \int (\boldsymbol{\omega} \wedge \mathbf{v}) \left(\mathbf{y}, t - \frac{|\mathbf{x}|}{c_0} \right) \cdot \nabla Y_j(\mathbf{y}) \, d^3\mathbf{y}, \quad |\mathbf{x}| \to \infty,$$

$$(7.3.3)$$

where

$$\mathbf{v} = U \nabla Y_1(\mathbf{y}).$$

There is no unsteady drag contribution to (7.3.3) from $j = 1$ because $\boldsymbol{\omega} \wedge \nabla Y_1 \cdot \nabla Y_1 \equiv 0$. Similarly, there can be no net side-force on the sphere because of the symmetric form of the vortex, and therefore there will be no contribution from $j = 3$. The sound is accordingly produced by a dipole source orientated in the x_2 direction; that is, the interaction produces an unsteady lift force in this direction which is responsible for the sound, which has the representation

$$p(\mathbf{x}, t) \approx \frac{-\rho_0 U \cos \Theta}{4\pi c_0 |\mathbf{x}|} \frac{\partial}{\partial t} \int (\boldsymbol{\omega} \cdot \nabla Y_1 \wedge \nabla Y_2) \left(\mathbf{y}, t - \frac{|\mathbf{x}|}{c_0} \right) d^3 \mathbf{y}, \quad |\mathbf{x}| \to \infty, \tag{7.3.4}$$

where $\Theta = \cos^{-1}(x_2/|\mathbf{x}|)$ is the angle between the observer direction $\mathbf{x}$ and the x_2 axis.

To evaluate the integral, write

$$\boldsymbol{\omega} = \Gamma \delta(\mathbf{s}_\perp) \hat{\mathbf{s}}, \qquad d^3 \mathbf{y} = d^2 \mathbf{s}_\perp \, ds,$$

where $\hat{\mathbf{s}}$ is a unit vector locally parallel to $\boldsymbol{\omega}$, $\mathbf{s}_\perp$ is the vector distance measured in the normal direction from the local axis of the vortex, and s is distance measured along the vortex in the direction of $\boldsymbol{\omega}$. The integral (7.3.4) can then be cast in the following nondimensional form, suitable for numerical evaluation,

$$\frac{p(\mathbf{x}, t)}{\rho_0 \Gamma U M \cos \Theta / 4\pi |\mathbf{x}|} = -\frac{\partial}{\partial T} \int_{-\infty}^{\infty} [\hat{\mathbf{s}} \cdot \nabla Y_1 \wedge \nabla Y_2] \, d\bar{s}, \quad \bar{s} = \frac{s}{a}, \quad M = \frac{U}{c_0}, \tag{7.3.5}$$

where the integrand is evaluated at the retarded position of the distorted vortex.

Now the integral in (7.3.5) is divergent, because $\hat{\mathbf{s}} \cdot \nabla Y_1 \wedge \nabla Y_2 \to 1$ as $\bar{s} \to \pm \infty$. The divergence is not real, however, but a consequence of the formal operations used in the application of the compact Green's function. The infinite contributions to the integral from large values of $\bar{s}$ are equal at successive retarded locations of the vortex, and disappear on differentiation with respect to T. The integral can therefore be evaluated numerically by restricting the range of integration to a finite interval, say, $-10 < \bar{s} < 10$, because the contributions at larger values of $\bar{s}$ are the same for all retarded times, and give no contribution to the sound when differentiated.

Typical plots of the calculated nondimensional pressure (7.3.5) are shown in Fig. 7.3.3 for two values of h/a; they illustrate how the sound level decreases as the initial standoff distance h of the vortex increases relative to the radius a of the sphere.

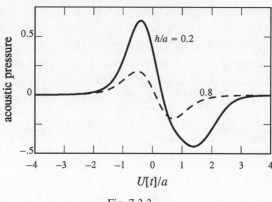

Fig. 7.3.3.

7.4 Compression Wave Generated When a Train Enters a Tunnel

A train entering a tunnel pushes aside the stationary air, most of which flows over the train and out of the tunnel portal, but the build-up of pressure just ahead of the train propagates into the tunnel as a compression wave at the speed of sound. In a long tunnel the compression wavefront can experience nonlinear steepening that is ultimately manifested as a loud, impulsive bang or 'crack' (called a *micro-pressure wave*) radiating out of the distant tunnel exit. In addition inaudible low-frequency pressure fluctuations called *infrasound* (at frequencies ~10–20 Hz) are radiated from the tunnel portal into the open air when the train enters and leaves the tunnel. All of these waves are indicated schematically in Fig. 7.4.1. Their effects become pronounced when the train speed U exceeds

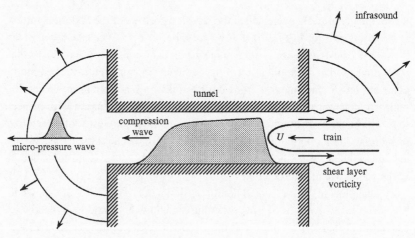

Fig. 7.4.1.

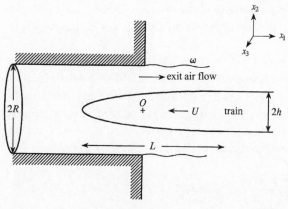

Fig. 7.4.2.

about 200 km/h (125 mi/h), when in particular the micro-pressure wave and the infrasound can cause vibrations and annoying structural rattles in neighboring buildings.

The formation of the compression wave can be studied in terms of the simpler problem involving an *axisymmetric* train entering axisymmetrically a semi-infinite, circular cylindrical duct of radius R and cross-sectional area $\mathcal{A} = \pi R^2$ (Fig. 7.4.2). Let the train be travelling at constant speed U in the negative x_1 direction, where the origin O is at the center of the tunnel entrance plane, so that the x_1 axis coincides with the axis of the tunnel. Denote the pressure, density, and speed of sound in the air respectively by $\bar{p}$, ρ and c. They vary with position and time within the tunnel, and their corresponding undisturbed values are p_0, ρ_0, c_0.

The cross section of the train is assumed to become uniform with constant area $\mathcal{A}_0 = \pi h^2$ at a distance L from the nose of the train, where h is the uniform maximum train radius. The aspect ratio h/L of the nose is taken to be sufficiently small, and the train profile sufficiently streamlined, to ensure that flow separation does not occur. In practice the Mach number $M = U/c_0$ does not exceed 0.4, and the *blockage* $\mathcal{A}_0/\mathcal{A} \leq 0.2$. If heat transfer and frictional losses are neglected during the initial stages of wave formation, the air flow may be regarded as *homentropic*, and the compression wave can be calculated using the vortex sound equation (5.2.5)

$$\left(\frac{D}{Dt} \left(\frac{1}{c^2} \frac{D}{Dt} \right) - \frac{1}{\rho} \nabla \cdot (\rho \nabla) \right) B = \frac{1}{\rho} \mathrm{div}(\rho \boldsymbol{\omega} \wedge \mathbf{v}). \qquad (7.4.1)$$

The air in the compression wave region ahead of the train may be regarded as linearly perturbed from its mean state, with $B \approx p/\rho_0 \equiv (\bar{p} - p_0)/\rho_0$. In this simple model the vorticity ω vanishes everywhere except within the outer shear layer of the exit flow of the air displaced when the train enters the tunnel (see Fig. 7.4.2).

Let $f \equiv f(x_1 + Ut, x_2, x_3) = 0$ be a control surface S contained within the fluid that just encloses the moving train, with $f < 0$ inside S (in the region occupied by the train) and $f > 0$ outside. The surface is fixed relative to the train, and the influence of the train on its surroundings can be represented in terms of monopole and dipole sources on S. In the usual way, multiply (7.4.1) by $H \equiv H(f)$ and rearrange (noting that $DH/Dt = 0$) to obtain

$$\left(\frac{D}{Dt} \left(\frac{1}{c^2} \frac{D}{Dt} \right) - \frac{1}{\rho} \nabla \cdot (\rho \nabla) \right) (HB)$$

$$= \frac{1}{\rho} \text{div}(H\rho\omega \wedge \mathbf{v}) - (\nabla B + \omega \wedge \mathbf{v}) \cdot \nabla H - \frac{1}{\rho} \text{div}(\rho B \nabla H). \quad (7.4.2)$$

This is a generalization of Equation (5.3.3). The two terms on the right-hand side involving ∇H respectively represent monopole and dipole sources distributed over the moving surface $f(x_1 + Ut, x_2, x_3) = 0$.

When frictional losses are neglected Crocco's equation (4.2.5) reduces to $\partial \mathbf{v}/\partial t = -\nabla B - \omega \wedge \mathbf{v}$, so that the source terms can be written

$$\frac{\partial}{\partial t} (\mathbf{U} \cdot \nabla H) - \mathbf{v} \cdot \nabla \frac{\partial H}{\partial t} - \frac{1}{\rho} \text{div} (\rho B \nabla H) + \frac{1}{\rho} \text{div}(H\rho\omega \wedge \mathbf{v}),$$

where $\mathbf{U} = (-U, 0, 0)$. The compressibility of the air adjacent to S and within the very low Mach number exterior flow from the tunnel portal can be neglected when $M(\mathcal{A}_0/\mathcal{A})^2 \ll 1$ (Howe et al. 2000), and the source approximated further by

$$\frac{\partial}{\partial t} (\mathbf{U} \cdot \nabla H) + \text{div} (\mathbf{v} \mathbf{U} \cdot \nabla H) - \text{div} \left\{ \left(\frac{p}{\rho_0} + \frac{1}{2} v^2 \right) \nabla H \right\} + \text{div}(H\omega \wedge \mathbf{v}),$$

where the relation $\partial H(f)/\partial t = -\mathbf{U} \cdot \nabla H(f)$ has been used.

Thus, if the nonlinear terms on the left of (7.4.2) (which affect the *propagation* of the compression wave) are also ignored, the equation finally reduces to

$$\left(\frac{1}{c_0^2} \frac{\partial^2}{\partial t^2} - \nabla^2 \right) (HB) = \frac{\partial}{\partial t} (\mathbf{U} \cdot \nabla H) + \text{div} (\mathbf{v} \mathbf{U} \cdot \nabla H)$$

$$- \text{div} \left\{ \left(\frac{p}{\rho_0} + \frac{1}{2} v^2 \right) \nabla H \right\} + \text{div}(H\omega \wedge \mathbf{v}). \quad (7.4.3)$$

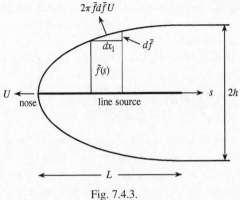

Fig. 7.4.3.

7.4.1 Linear Theory

When the blockage $\mathcal{A}_0/\mathcal{A}$ is small it is sufficient to retain only the first *monopole* source on the right-hand side of Equation (7.4.3). This source can be simplified when the aspect ratio $h/L \ll 1$ by introducing a *slender body* approximation. To do this, consider a system of cylindrical coordinates in which $r = \sqrt{x_2^2 + x_3^2}$ is the perpendicular distance from the axis of the train. The control surface equation

$$f(x_1 + Ut, x_2, x_3) = 0 \quad \text{can be written } r = \bar{f}(x_1 + Ut),$$

where $\mathcal{A}_T(s) = \pi \bar{f}^2(s)$ is the cross-sectional area of the train at distance s from the nose, and the nose is assumed to cross the tunnel entrance plane $(x_1 = 0)$ at $t = 0$. As the train moves (to the left in Fig. 7.4.3) the rate at which air is displaced by a section of the train of length dx_1 is

$$U 2\pi \bar{f}(x_1 + Ut) \, d\bar{f}(x_1 + Ut) = U 2\pi \bar{f}(x_1 + Ut) \frac{\partial \bar{f}}{\partial x_1}(x_1 + Ut) \, dx_1$$

$$\equiv U \frac{\partial \mathcal{A}_T}{\partial x_1}(x_1 + Ut) \, dx_1.$$

Thus,

$$U \frac{\partial \mathcal{A}_T}{\partial x_1}(x_1 + Ut) = \text{monopole source strength per unit length of the train.}$$

When $h/L \ll 1$, we can collapse the monopole source distribution over the surface of the train into a *line source* concentrated on its axis, and approximate the monopole on the right-hand side of Equation (7.4.3) by

$$\frac{\partial}{\partial t}(\mathbf{U} \cdot \nabla H)(\mathbf{x}, t) \approx \frac{\partial}{\partial t}\left(U \frac{\partial \mathcal{A}_T}{\partial x_1}(x_1 + Ut)\delta(x_2)\delta(x_3)\right).$$

The source strength is proportional to the rate at which the train cross section changes with distance along the train, and is nonzero only in the vicinity of the train nose (and also the tail).

The corresponding approximation of Equation (7.4.3) is therefore

$$\left(\frac{1}{c_0^2}\frac{\partial^2}{\partial t^2} - \nabla^2\right)(HB) = \frac{\partial}{\partial t}\left(U\frac{\partial \mathcal{A}_T}{\partial x_1}(x_1 + Ut)\delta(x_2)\delta(x_3)\right), \quad (7.4.4)$$

where $B \to p/\rho_0$ in the linear acoustic region ahead of the train.

The monopole in this equation does not depend on time when viewed in a reference frame moving at the uniform *subsonic* speed U of the train. The source therefore creates only a *nonacoustic* near field when travelling within the tunnel or in free space far from the tunnel entrance (c.f., Question 4 of Problems 1 when $q_0(t) = \text{constant}$). The compression wave is produced when the near field of the source interacts with the tunnel portal, as the train nose enters the tunnel. This occurs over a time $\sim R/U$, so that the characteristic thickness of the wavefront $\sim R/M \gg R$. Equation (7.4.4) can therefore be solved by using the compact Green's function (3.9.13) for a duct entrance

$$G(\mathbf{x}, \mathbf{y}; t - \tau) \approx \frac{c_0}{2\mathcal{A}}\left\{H\left(t - \tau - \frac{|\varphi^*(\mathbf{x}) - \varphi^*(\mathbf{y})|}{c_0}\right)\right.$$
$$\left. - H\left(t - \tau + \frac{\varphi^*(\mathbf{x}) + \varphi^*(\mathbf{y})}{c_0}\right)\right\},$$

where $\varphi^*(\mathbf{x})$ is the velocity potential of a uniform incompressible flow *out* of the tunnel portal that has unit speed far inside the tunnel (see (3.9.14)).

Thus, at $\mathbf{x}$ within the tunnel, ahead of the train where $B = p/\rho_0$, we have

$$p \equiv p\left(t + \frac{x_1}{c_0}\right) \approx \rho_0\frac{\partial}{\partial t}\iint_{-\infty}^{\infty} U\frac{\partial \mathcal{A}_T}{\partial y_1}(y_1 + U\tau)G(\mathbf{x}, y_1, 0, 0; t - \tau)\,dy_1\,d\tau$$

$$= \frac{\rho_0 U c_0}{2\mathcal{A}}\int_{-\infty}^{\infty}\{\mathcal{A}_T'(y_1 - M\varphi^*(y_1, 0, 0) + U[t])$$
$$- \mathcal{A}_T'(y_1 + M\varphi^*(y_1, 0, 0) + U[t])\}\,dy_1, \quad (7.4.5)$$

where the prime on $\mathcal{A}_T$ denotes differentiation with respect to the argument, and $[t] = t + (x_1 - \ell')/c_0$ is the effective retarded time. Because nonlinear propagation terms have been ignored, this approximation determines the *initial* form of the compression wave profile, before the onset of nonlinear steepening. It is therefore applicable within the region several tunnel diameters ahead of the train, during and just after tunnel entry.

The main contributions to the Integral (7.4.5) are from the vicinities of the nose and tail of the train, where the cross-sectional area $\mathcal{A}_T$ is changing. The compression wave is generated as the nose enters the tunnel, and may be calculated by temporarily considering a train of semi-infinite length. During the formation of the wave, and in the particular case in which the Mach number is small enough that terms $\sim O(M^2)$ are negligible, the term $M\varphi^*$ in the arguments of $\mathcal{A}'_T$ in (7.4.5) is small, and we then find, by expanding to first order in $M\varphi^*$ and integrating by parts, that

$$p \approx \frac{\rho_0 U^2}{\mathcal{A}} \int_{-\infty}^{\infty} \frac{\partial \mathcal{A}_T}{\partial y_1}(y_1 + U[t]) \frac{\partial \varphi^*}{\partial y_1}(y_1, 0, 0)\, dy_1, \qquad M^2 \ll 1. \quad (7.4.6)$$

After the nose has passed into the tunnel, $\partial \varphi^* / \partial y_1 = 1$ in the region occupied by the nose, and (7.4.6) predicts the overall (linear theory) pressure rise to be $\Delta p \approx \rho_0 U^2 \mathcal{A}_0 / \mathcal{A}$. But the linear theory, asymptotic pressure rise can also be calculated *exactly*, with no restriction on Mach number, to be

$$\Delta p = \frac{\rho_0 U^2 \mathcal{A}_0}{\mathcal{A}(1 - M^2)},$$

because this is attained when $\varphi^*(y_1, 0, 0) \approx y_1 - \ell'$ in (7.4.5) (see Equation (3.9.14)). This implies that the Approximation (7.4.6) can be extrapolated to finite Mach numbers by writing

$$p \approx \frac{\rho_0 U^2}{\mathcal{A}(1 - M^2)} \int_{-\infty}^{\infty} \frac{\partial \mathcal{A}_T}{\partial y_1}(y_1 + U[t]) \frac{\partial \varphi^*}{\partial y_1}(y_1, 0, 0)\, dy_1. \quad (7.4.7)$$

This extrapolation of the linear theory to finite values of M turns out to be applicable for $M < 0.4$ (Howe et al. 2000).

Figure 7.4.4 illustrates schematically an experimental arrangement used by Maeda et al. (1993) to investigate the compression wave. Wire-guided, axisymmetric model trains are projected into and along the axis of a tunnel consisting of a 7-m long circular cylinder of internal diameter 0.147 m. The nose aspect ratio $h/L = 0.2$, the blockage $\mathcal{A}_0/\mathcal{A} = 0.116$, and the projection speed $U \approx 230$ km/h ($M \approx 0.188$). The train nose profiles include the cone, and the paraboloid and ellipsoid of revolution, with respective cross-sectional areas given by

$$\frac{\mathcal{A}_T(s)}{\mathcal{A}_0} = \begin{cases} \frac{s^2}{L^2},\ \frac{s}{L},\ \frac{s}{L}\left(2 - \frac{s}{L}\right), & 0 < s < L, \\ 1, & s \geq L. \end{cases}$$

The data points in the figure are measurements (made at the point labelled T) of the pressure gradient dp/dt 1 m from the entrance for these three different

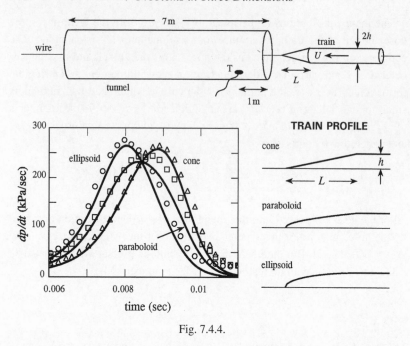

Fig. 7.4.4.

nose profiles. The solid curves are predictions of Equation (7.4.7), evaluated using the following formulae for $\partial\varphi^*/\partial y_1$ (Howe 1998b):

$$\frac{\partial\varphi^*}{\partial y_1}(\mathbf{y}) = \frac{1}{2} - \frac{1}{2\pi}\int_0^\infty I_0\left(\frac{\xi r}{R}\right)\left(\frac{2K_1(\xi)}{I_1(\xi)}\right)^{\frac{1}{2}}\sin\left\{\xi\left(\frac{y_1}{R} + \mathcal{Z}(\xi)\right)\right\}d\xi,$$

$$\mathcal{Z}(\xi) = \frac{1}{\pi}\int_0^\infty \ln\left(\frac{K_1(\mu)I_1(\mu)}{K_1(\xi)I_1(\xi)}\right)\frac{d\mu}{\mu^2 - \xi^2},$$

where $r = \sqrt{y_2^2 + y_3^2} < R$ and I_0, I_1, and K_1 are modified Bessel functions.

The linear theory underpredicts the maximum observed pressure gradients by about 8%. The agreement with experiment can be greatly improved by including contributions from the surface dipoles in Equation (7.4.3) (which in a first approximation are determined by the drag force exerted on the nose of the train by the linear theory pressure rise) and, to a lesser extent, by including the vortex sound generated by the tunnel exit-flow vorticity (the final source term on the right of (7.4.3)).

Problems 7

1. The term $\omega_2 \partial Y_2/\partial y_3$ in the Representation (7.1.3) of the sound produced by a gust interacting with a thin airfoil accounts for the influence of changes in

the airfoil chord $2a(y_3)$ over the interaction region. Show that when

$$\frac{da}{dy_3}(y_3) \ll 1$$

the Formula (7.2.3) for the sound produced by a line vortex is given in a first approximation by

$$p(\mathbf{x}, t) \approx \frac{\rho_0 \Gamma U M \cos \Theta}{2^{\frac{3}{2}} \pi |\mathbf{x}|} \frac{|\sin \phi|}{\sin \theta}$$
$$\times \left\{ \cos \theta - \sin \theta \cos \phi \left(\frac{da}{dy_3} \right)_0 \right\} \frac{H\left(\frac{U[t]}{a} + 1 \right)}{\sqrt{\frac{U[t]}{a} + 1}},$$

where da/dy_3 is evaluated at $y_3 = 0$, where the vortex is cut by the airfoil.

Show that this result is identical with that given by (7.1.4) provided that in (7.1.4) ω_3 is interpreted as the component of the vorticity parallel to the local leading edge of the airfoil and the convection velocity U is replaced by its component normal to the local leading edge.

2. A vortex ring orientated with its axis parallel to the $+x_1$ axis is convected in a low Mach number mean flow at speed U in the x_1 direction past the edge of the rigid half-plane $x_1 < 0$, $x_2 = 0$, $-\infty < x_3 < \infty$. Use the compact Green's function

$$G(\mathbf{x}, \mathbf{y}, t - \tau) = \frac{1}{2\pi^2 \sqrt{2\pi i c_0}} \frac{\varphi^*(\mathbf{x})\varphi^*(\mathbf{y})}{|\mathbf{x} - y_3\mathbf{i}_3|^{3/2}} \int_{-\infty}^{\infty} \sqrt{\omega} e^{-i\omega(t-\tau-|\mathbf{x}-y_3\mathbf{i}_3|/c_0)} \, d\omega,$$

where φ^* is defined as in (3.9.6), to calculate the sound produced as the vortex passes the edge when the influence of the half-plane on the motion of the ring is ignored.

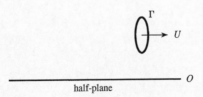

half-plane

3. Calculate the sound produced within and outside a semi-finite circular cylindrical rigid pipe when a vortex ring exhausts axisymmetrically from the open end. Neglect the influence of the pipe walls on the motion of the vortex and ignore any change in the vortex radius at the exit.

4. Determine the (quadrupole) sound produced by the head-on collision of two equal ring vortices. Estimate the sound generated when a ring vortex is incident normally on a plane wall.

5. Use the Green's function (3.9.15) and Equation (7.4.4) to determine the infrasound generated by a train entering a tunnel modeled by the unflanged, circular cylindrical duct in Fig. 3.9.6b. Assume that the train travels along the axis of the duct and show that the acoustic pressure at the far field point **x** outside the tunnel is given approximately by

$$p(\mathbf{x}, t) \approx \frac{\rho_0 U^2 M}{4\pi |\mathbf{x}|} \left(1 - \frac{x_1}{|\mathbf{x}|} \right)$$

$$\times \int_{-\infty}^{\infty} \frac{\partial \mathcal{A}_T}{\partial y_1} (y_1 + U[t]) \frac{\partial^2 \varphi^*}{\partial y_1^2} (y_1, 0, 0) \, dy_1, \quad |\mathbf{x}| \to \infty,$$

where $[t] = t - |\mathbf{x}|/c_0$.

8

Further Worked Examples

8.1 Blade–Vortex Interactions in Two Dimensions

The linear theory of the low Mach number, two-dimensional interaction of a line vortex with an airfoil was discussed in Section 6.4. The interaction will now be examined in more detail, including also the influence of *image* vortices on the motion. The general problem to be considered is depicted in Fig. 8.1.1, which shows a vortex of strength Γ moving in the neighbourhood of a rigid airfoil of chord $2a$ occupying $-a < x_1 < a$, $x_2 = 0$. There is no mean circulation about the airfoil. We shall consider cases with and without a mean flow in the x_1 direction and examine the influence of vortex shedding from the trailing edge.

8.1.1 Equation of Motion of the Vortex

At time t let the vortex be at

$$\mathbf{x} \equiv (x_1, x_2) = \mathbf{x}_0(t), \quad \text{and translate at velocity } \mathbf{v}_0 = \frac{d\mathbf{x}_0}{dt}(t).$$

If we set $z = x_1 + i x_2$, $z_0 = x_{01} + i x_{02}$, the transformation

$$\zeta = \frac{z}{a} + \sqrt{\frac{z^2}{a^2} - 1} \tag{8.1.1}$$

maps the fluid region in the z plane of the airfoil into the region $|\zeta| > 1$ in the ζ plane. The upper and lower faces of the airfoil ($x_2 = \pm 0$) respectively transform into the upper and lower halves of the unit circular cylinder $|\zeta| = 1$, and the vortex maps into an equal vortex at $\zeta = \zeta_0$ (Fig. 8.1.2). In the absence of mean flow ($U = 0$), and when there is no mean circulation about the cylinder (and therefore about the airfoil), the complex potential of the motion is obtained by placing an image vortex $-\Gamma$ at the *inverse point* $\zeta = 1/\zeta_0^*$ together with

175

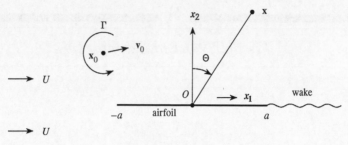

Fig. 8.1.1.

a vortex $+\Gamma$ at the center of the cylinder. The two interior vortices ensure that the total circulation around the cylinder vanishes. Then,

$$w(\zeta) = -\frac{i\Gamma}{2\pi}\ln(\zeta - \zeta_0) + \frac{i\Gamma}{2\pi}\ln\left(\zeta - \frac{1}{\zeta_0^*}\right) - \frac{i\Gamma}{2\pi}\ln\zeta.$$

The velocity potential of the motion in the z plane is given by setting $\zeta = \zeta(z)$. Because a mean flow in the x_1 direction is unaffected by the airfoil, we can include its contribution by adding the complex potential Uz. Then,

$$w(z) = -\frac{i\Gamma}{2\pi}\ln(\zeta(z) - \zeta(z_0)) + F(z),$$

where $\quad F(z) = \frac{i\Gamma}{2\pi}\ln\left(\zeta(z) - \frac{1}{\zeta(z_0)^*}\right) - \frac{i\Gamma}{2\pi}\ln\zeta(z) + Uz.$

This is of the form given in (4.6.1), so that the corresponding equation of motion

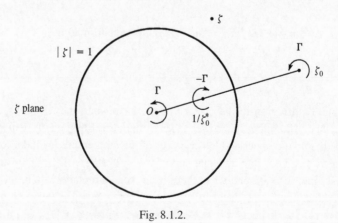

Fig. 8.1.2.

of the vortex at z_0 is found from (4.6.3) to be

$$\frac{dz_0^*}{dt} \equiv \frac{dx_{01}}{dt} - i\frac{dx_{02}}{dt} = -\frac{i\Gamma \zeta''(z_0)}{4\pi \zeta'(z_0)} + F'(z_0),$$

that is
$$\frac{dz_0^*}{dt} = \frac{i\Gamma}{4\pi a\sqrt{Z^2 - 1}} \left\{ \frac{Z}{\sqrt{Z^2 - 1}} - 1 + \frac{2}{|\zeta_0|^2 - 1} \right\} + U, \quad (8.1.2)$$

where $\quad Z = \dfrac{z_0}{a}, \qquad \zeta_0 = Z + \sqrt{Z^2 - 1}.$

Equation (8.1.2) takes no account of the influence of vortex shedding. In a linearized calculation (in which image effects in the airfoil are neglected) this could be done by assuming shed vorticity to lie in a thin vortex sheet downstream of the trailing edge at $x_1 = a$, as in Section 6.3. This would lead to a solution in terms of the Sears function (6.3.6), but we shall not do this, because it limits the discussion to linearized motions. Instead, we shall apply the method discussed in Section 6.4, where the effects of vortex shedding are modelled by deleting singularities from the compact Green's function.

8.1.2 Formula for the Acoustic Pressure

The sound produced by the low Mach number motion of the vortex is calculated from (6.2.4):

$$p(\mathbf{x}, t) \approx \frac{-\rho_0 \Gamma x_j}{2\pi \sqrt{2c_0}|\mathbf{x}|^{\frac{3}{2}}} \frac{\partial}{\partial t} \int_{-\infty}^{t-|\mathbf{x}|/c_0} \left\{ \frac{dx_{01}}{d\tau} \frac{\partial Y_j}{\partial y_2} - \frac{dx_{02}}{d\tau} \frac{\partial Y_j}{\partial y_1} \right\}_{\mathbf{x}_0(\tau)}$$

$$\times \frac{d\tau}{\sqrt{t - \tau - |\mathbf{x}|/c_0}}, \qquad (8.1.3)$$

where the Kirchhoff vector for the strip airfoil has the components (Table 3.9.1)

$$Y_1 = y_1, \qquad Y_2 = \mathrm{Re}(-i\sqrt{z^2 - a^2}), \qquad z = y_1 + iy_2. \quad (8.1.4)$$

By defining the radiation angle Θ for an observer at $\mathbf{x}$ in the far field as in Fig. 8.1.1, we can write

$$p(\mathbf{x}, t) \approx \frac{\rho_0 \Gamma \sin \Theta}{2\pi \sqrt{2c_0|\mathbf{x}|}} \frac{\partial}{\partial t} \int_{-\infty}^{t-|\mathbf{x}|/c_0} \frac{dx_{02}}{d\tau}(\tau) \frac{d\tau}{\sqrt{t - \tau - |\mathbf{x}|/c_0}} - \frac{\rho_0 \Gamma \cos \Theta}{2\pi \sqrt{2c_0|\mathbf{x}|}}$$

$$\times \frac{\partial}{\partial t} \int_{-\infty}^{t-|\mathbf{x}|/c_0} \left(\frac{dx_{01}}{d\tau} \frac{\partial Y_2}{\partial y_2} - \frac{dx_{02}}{d\tau} \frac{\partial Y_2}{\partial y_1} \right)_{\mathbf{x}_0(\tau)} \frac{d\tau}{\sqrt{t - \tau - |\mathbf{x}|/c_0}}.$$

The two integrals in this formula represent the acoustic fields of dipole sources. According to Curle's theory (Section 2.3) and the Formula (6.2.2), the strengths of these dipoles are determined by the unsteady force (F_1, F_2) exerted on the fluid (per unit span) by the airfoil. The first is aligned with the airfoil chord (the mean flow direction) and represents the influence of *suction* forces at the leading and trailing edges (Batchelor 1967); the second component F_2 is equal and opposite to the unsteady lift experienced by the airfoil during the interaction.

In general the integrals must be evaluated numerically using the solution of the equation of motion (8.1.2). Introduce the shorthand notation

$$\mathcal{W} = \frac{d}{dz}(-i\sqrt{z^2 - a^2}) = \frac{-iZ}{\sqrt{Z^2 - 1}} \qquad (8.1.5)$$

evaluated at the vortex. Then,

$$\left(\frac{dx_{01}}{d\tau}\frac{\partial Y_2}{\partial y_2} - \frac{dx_{02}}{d\tau}\frac{\partial Y_2}{\partial y_1}\right)_{\mathbf{x}_0(\tau)} \equiv -a\,\mathrm{Im}\left(\mathcal{W}(Z)\frac{dZ}{d\tau}\right),$$

and the acoustic pressure becomes

$$p(\mathbf{x}, t) \approx \frac{\rho_0\Gamma a}{2\pi\sqrt{2c_0|\mathbf{x}|}}\frac{\partial}{\partial t}\left\{\sin\Theta\int_{-\infty}^{[t]}\mathrm{Im}\left(\frac{dZ}{d\tau}\right)\frac{d\tau}{\sqrt{[t] - \tau}}\right.$$
$$\left. + \cos\Theta\int_{-\infty}^{[t]}\mathrm{Im}\left(\mathcal{W}(Z)\frac{dZ}{d\tau}\right)\frac{d\tau}{\sqrt{[t] - \tau}}\right\}, \quad (8.1.6)$$

where $[t] = t - |\mathbf{x}|/c_0$ is the retarded time, and it is understood that $Z = Z(\tau)$.

Equations (8.1.2) and (8.1.6) for the vortex motion and the acoustic pressure will now be applied to several different special cases.

8.1.3 Linear Theory

In the linearized approximation the vortex is swept past the airfoil along a trajectory parallel to the x_1 direction at precisely the uniform mean flow speed U. This is the case illustrated in Fig. 6.4.1. When the standoff distance $h \ll a$ it was argued in Section 6.4 that the influence of vortex shedding from the trailing edge could be estimated by deleting the singularity that occurs at the edge from the Green's function and ignoring the shed vorticity. Only the second integral in (8.1.6) contributes to the sound (because $dZ/d\tau = U/a$ is real and $F_1 \equiv 0$), and the trailing edge singularity corresponds to the singularity of $\mathcal{W}(Z)$ at $Z = 1$. $\mathcal{W}(Z)$ is singular at both the leading and trailing edges ($Z = \pm 1$), which, other things being equal, are therefore the most significant sources of sound at high frequencies, because the second integral in (8.1.6) is dominated by contributions from the neighbourhoods of the singularities. By deleting the contribution from the trailing edge we are asserting that all of the sound is produced by the interaction of the vortex with the leading edge. Near this edge

$$W(Z) \approx \frac{1}{\sqrt{2}\sqrt{Z+1}}, \tag{8.1.7}$$

where the branch cut for $\sqrt{Z+1}$ runs along the real axis from $Z = -1$ to $Z = +\infty$. Making this substitution in (8.1.6), measuring time from the instant that the vortex crosses the midchord $x_1 = 0$ of the airfoil, so that $Z = U\tau/a + ih/a$, and changing the integration variable to $\xi = 1/\sqrt{t - \tau - |\mathbf{x}|/c_0}$, we then recover the result (6.4.3), which can be written,

$$\frac{p(\mathbf{x}, t)}{\rho_0 \Gamma U \sqrt{M} \cos \Theta (a/|\mathbf{x}|)^{\frac{1}{2}}/4\pi a} \approx \frac{\left(\frac{U[t]}{a} + 1\right)}{\left(\frac{U[t]}{a} + 1\right)^2 + \left(\frac{h}{a}\right)^2}, \quad |\mathbf{x}| \to \infty. \tag{8.1.8}$$

The nondimensional acoustic pressure signature (the right-hand side of (8.1.8)) is plotted as the solid curves in Fig. 8.1.3 for $h/a = 0.2, 0.5, 1.0$.

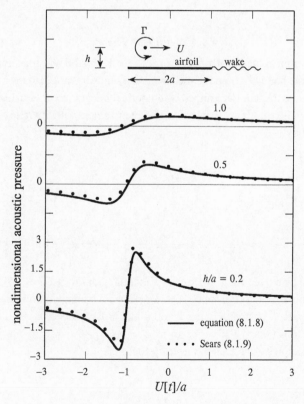

Fig. 8.1.3.

It was pointed out in Section 6.3 that the linearized problem of determining the unsteady force F_2 exerted on the fluid when an incompressible, sinusoidal gust convects past the airfoil can be solved exactly with full account taken of vortex shedding, in terms of the Sears function (6.3.6). This force also determines the low Mach number acoustic radiation by Equation (6.2.2) (because $F_1 \equiv 0$), which in the present case can be shown to predict that

$$\frac{p(\mathbf{x}, t)}{\rho_0 \Gamma U \sqrt{M} \cos \Theta (a/|\mathbf{x}|)^{\frac{1}{2}} / 4\pi a} \approx \sqrt{2\pi} \mathrm{Re} \int_0^\infty (i\lambda)^{\frac{1}{2}} S(\lambda) e^{-\lambda \{ h/a + iU[t]/a \}} \, d\lambda,$$

$$|\mathbf{x}| \to \infty. \qquad (8.1.9)$$

The corresponding pressure signatures are plotted as the dotted curves in Fig. 8.1.3. The agreement with the approximate theory of Equation (8.1.8) is remarkably good even when h/a is as large as unity, when the characteristic reduced frequency $\lambda = \omega a/U$ of the motion is relatively small, and might be expected to lie outside the range for which (8.1.8) is valid.

8.1.4 Nonlinear Theory

When account is taken of image vortices in the airfoil the trajectory of the vortex in the neighbourhood of the airfoil is no longer parallel to the mean flow direction, and must be determined by numerical integration of Equation (8.1.2). To do this it is convenient to introduce a dimensionless velocity ratio ϵ and time T defined by

$$\epsilon = \frac{\Gamma}{4\pi a U}, \qquad T = \frac{Ut}{a},$$

in terms of which (8.1.2) becomes

$$\frac{dZ^*}{dT} = \frac{i\epsilon}{\sqrt{Z^2 - 1}} \left\{ \frac{Z}{\sqrt{Z^2 - 1}} - 1 + \frac{2}{|\zeta_0|^2 - 1} \right\} + 1, \qquad (8.1.10)$$

which can be solved for Z by Runge–Kutta integration (Section 4.6).

If the initial standoff distance is h at $x_1 = -\infty$, the integration is started at a large distance L upstream of the airfoil midchord, say $L = 10a$, by prescribing the initial position of the vortex to be $Z = -L/a + ih/a$. The upper part of Fig. 8.1.4 shows a calculated trajectory for

$$\epsilon = 0.2, \qquad \frac{h}{a} = 0.2,$$

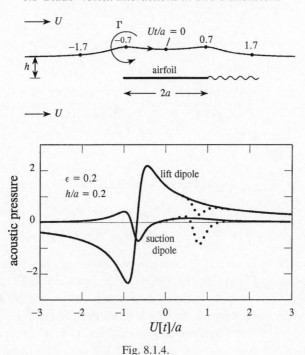

Fig. 8.1.4.

where time is measured from the instant that the vortex passes the midchord of the airfoil. The nonlinear influence of the image vorticity is to shift the initially rectilinear trajectory of the vortex away from the airfoil in the direction of the vortex force $\omega \wedge \mathbf{U}$ ($\mathbf{U} = U\mathbf{i}$). The vortex is closest to the airfoil at $Ut/a = 0$, where $x_{02} \sim 0.28a$, and where convection by the images increases the translation speed of the vortex from U to approximately

$$U + \frac{\Gamma}{4\pi x_{02}} = U + \frac{\epsilon U}{x_{02}/a} \sim U\left(1 + \frac{0.2}{0.28}\right) = 1.71U.$$

The sound generated as the vortex passes the airfoil is given by (8.1.6). The influence of vortex shedding into the wake is included by using the approximation (8.1.7) for $\mathcal{W}(Z)$. The integrals must be evaluated numerically, and this is done by defining a dimensionless vortex convection velocity $(u(\hat{T}), v(\hat{T}))$ by

$$\frac{dZ}{d\hat{T}} = u(\hat{T}) + iv(\hat{T}), \quad \text{where } \hat{T} = \frac{U\tau}{a}.$$

Then,

$$
\frac{p(\mathbf{x}, t)}{\rho_0 \Gamma U \sqrt{M}(a/|\mathbf{x}|)^{\frac{1}{2}}\big/ 4\pi a}
$$

$$
\approx 2^{\frac{1}{2}} \frac{d}{dT} \left\{ \sin\Theta \int_{-\infty}^{[T]} \frac{v(\hat{T})\, d\hat{T}}{\sqrt{[T]-\hat{T}}} + \cos\Theta \int_{-\infty}^{[T]} \frac{\mathrm{Im}(\mathcal{W}(Z)(u+iv))(\hat{T})\, d\hat{T}}{\sqrt{[T]-\hat{T}}} \right\}
$$

$$
= 2^{\frac{3}{2}} \sin\Theta \frac{d}{dT} \int_0^\infty v([T]-\lambda^2)\, d\lambda
$$

$$
+ 2^{\frac{3}{2}} \cos\Theta \frac{d}{dT} \int_0^\infty \mathrm{Im}(\mathcal{W}(Z)(u+iv))([T]-\lambda^2)\, d\lambda
$$

$$
\approx \frac{p_1(\mathbf{x}, t)}{\rho_0 \Gamma U \sqrt{M}(a/|\mathbf{x}|)^{\frac{1}{2}}\big/ 4\pi a} + \frac{p_2(\mathbf{x}, t)}{\rho_0 \Gamma U \sqrt{M}(a/|\mathbf{x}|)^{\frac{1}{2}}\big/ 4\pi a}, \tag{8.1.11}
$$

where $T = Ut/a$, $[T] = U[t]/a$, and the integration variable $\hat{T}$ has been replaced by $\lambda = \sqrt{[T]-\hat{T}}$. The final integrals are easily evaluated numerically when the path of the vortex has been determined. The upper limit of integration is actually finite, because the source terms must be set to equal zero as soon as $[T] - \lambda^2$ reduces to the nondimensional time at which the computation of the vortex path begins (where the vortex is sufficiently far upstream that it effectively produces no sound by interaction with the airfoil).

The components $p_1(\mathbf{x}, t)$, $p_2(\mathbf{x}, t)$ of (8.1.11) correspond respectively to the dipole sound produced by the unsteady suction and lift forces; their nondimensional forms

$$
\frac{p_1(\mathbf{x}, t)}{\rho_0 \Gamma U \sqrt{M} \sin\Theta (a/|\mathbf{x}|)^{\frac{1}{2}}\big/ 4\pi a} \quad \text{and} \quad \frac{p_2(\mathbf{x}, t)}{\rho_0 \Gamma U \sqrt{M} \cos\Theta (a/|\mathbf{x}|)^{\frac{1}{2}}\big/ 4\pi a}
$$

are plotted in Fig. 8.1.4. Vortex shedding should smooth out the pressure signatures at the retarded times when the vortex is close to the trailing edge. But the calculated pressures exhibit blips shown as dotted curves in the figure. These arise because, although our calculation has accounted for vortex shedding in evaluating the dipole source strengths (by means of the approximation (8.1.7)), the effect of shedding was *not* included in the calculation of the vortex *trajectory*. However, the smoothing influence of shedding at a sharp edge acts to remove the blips, and the pressure signatures have profiles similar to those depicted by the solid curves in the figure, obtained by interpolating smoothly between the calculated pressures on either side of the blips.

An interesting nonlinear interaction occurs when the initial standoff distance of the vortex $h = 0$ (Fig. 8.1.5). In the linearized approximation, the vortex would strike the leading edge of the airfoil at $U[t]/a = -1$, at which time the

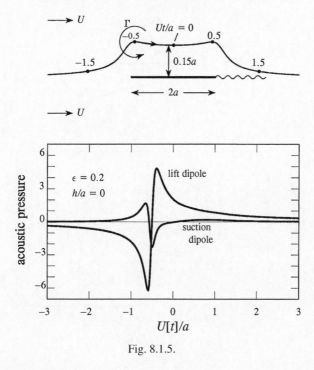

Fig. 8.1.5.

linear theory acoustic pressure (8.1.8) is infinite. This singular event does not occur because the vortex trajectory is deflected around the airfoil by the image vorticity (for a rounded nose the possibility of additional vortex shedding from the leading edge may be ignored). The upper part of Fig. 8.1.5 illustrates this for the same value of the velocity ratio $\epsilon = \Gamma/4\pi aU = 0.2$ considered above. The maximum convection velocity of the vortex (at $Ut/a = 0$) is now more than twice the mean stream velocity:

$$U + \frac{\Gamma}{4\pi(0.15a)} = U\left(1 + \frac{0.2}{0.15}\right) = 2.3U.$$

The corresponding suction- and lift-dipole acoustic pressures p_1 and p_2 shown in the figure are also greatly increased.

8.1.5 Periodic Vortex Motion

When there is no mean flow ($U = 0$) the characteristic velocity and dimensionless time become

$$V = \frac{\Gamma}{4\pi a}, \qquad T = \frac{Vt}{a},$$

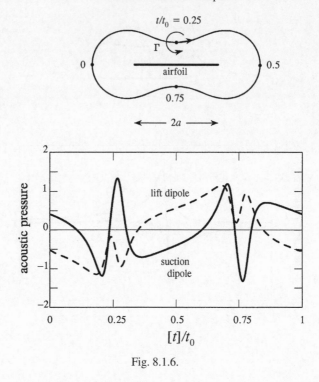

Fig. 8.1.6.

and the vortex equation of motion (8.1.2) reduces to

$$\frac{dZ^*}{dT} = \frac{i}{\sqrt{Z^2-1}}\left\{\frac{Z}{\sqrt{Z^2-1}} - 1 + \frac{2}{|\zeta_0|^2-1}\right\}.$$

The solutions are closed trajectories orbiting the airfoil periodically. A typical orbit is plotted in the upper half of Fig. 8.1.6 for $\Gamma > 0$, for the case where the trajectory passes through the point labelled 0, where $x_{01} = -2a$, $x_{02} = 0$. The calculated period is $T_0 \equiv Vt_0/a \approx 35.84$.

An orbiting vortex motion of this kind cannot be realized in practice (because of diffusion from the vortex core and the continual shedding of additional vorticity from the airfoil), but it is still instructive to calculate the sound produced by the motion. By writing

$$\hat{T} = \frac{V\tau}{a} \quad \text{and} \quad \frac{dZ}{d\hat{T}} = u(\hat{T}) + iv(\hat{T})$$

in the general formula (8.1.6) for the acoustic pressure, the nondimensional suction and lift acoustic pressures are found to be given by the following modified form of (8.1.11)

$$
\frac{p_1(\mathbf{x}, t)}{\rho_0 V^2 \sqrt{M}(a/|\mathbf{x}|)^{\frac{1}{2}}} + \frac{p_2(\mathbf{x}, t)}{\rho_0 V^2 \sqrt{M}(a/|\mathbf{x}|)^{\frac{1}{2}}}
$$

$$
\approx 2^{\frac{1}{2}} \frac{d}{dT} \left\{ \sin \Theta \int_{-\infty}^{[T]} \frac{v(\hat{T}) \, d\hat{T}}{\sqrt{[T] - \hat{T}}} + \cos \Theta \int_{-\infty}^{[T]} \frac{\mathrm{Im}(\mathcal{W}(Z)(u + iv))(\hat{T}) \, d\hat{T}}{\sqrt{[T] - \hat{T}}} \right\},
$$
(8.1.12)

where $M = V/c_0$.

The function $\mathcal{W}(Z)$ in the integrand is given by (8.1.5) in the absence of vortex shedding. It follows by inspection and from the numerical solution, that when T is measured from $Z = -2$, as indicated in Fig. 8.1.6, the suction and dipole source strengths have period T_0, and possess Fourier series expansions of the form

$$
v(T) = \sum_{n=1}^{\infty} a_n \cos \left(\frac{2\pi n T}{T_0} \right), \quad \mathrm{Im}(\mathcal{W}(Z)(u+iv))(T) = \sum_{n=1}^{\infty} b_n \sin \left(\frac{2\pi n T}{T_0} \right),
$$

where the coefficients a_n, b_n can be calculated by using the numerical solution for the orbit to evaluate

$$
a_n = \frac{2}{T_0} \int_0^{T_0} v(T) \cos \left(\frac{2\pi n T}{T_0} \right) dT,
$$

$$
b_n = \frac{2}{T_0} \int_0^{T_0} \mathrm{Im}(\mathcal{W}(Z)(u + iv))(T) \sin \left(\frac{2\pi n T}{T_0} \right) dT.
$$

By making the change of integration variable $\lambda = \sqrt{[T] - \hat{T}}/\sqrt{T_0}$, the right-hand side of (8.1.12) now becomes

$$
\frac{4\sqrt{2\pi}}{\sqrt{T_0}} \sum_{n=1}^{\infty} \left\{ -a_n n \sin \Theta \int_0^{\infty} \sin \left[2\pi n \left(\frac{[T]}{T_0} - \lambda^2 \right) \right] d\lambda \right.
$$

$$
\left. + b_n n \cos \Theta \int_0^{\infty} \cos \left[2\pi n \left(\frac{[T]}{T_0} - \lambda^2 \right) \right] d\lambda \right\}.
$$

The integrals are evaluated from the real and imaginary parts of

$$
\int_0^{\infty} e^{2\pi i n \{[T]/T_0 - \lambda^2\}} \, d\lambda = \frac{1}{2\sqrt{2n}} e^{\{2n[T]/T_0 - \frac{1}{4}\} \pi i}.
$$

Hence, the suction and lift force dipole fields are given respectively by

$$\frac{p_1(\mathbf{x}, t)}{\rho_0 V^2 \sqrt{M} \sin \Theta (a/|\mathbf{x}|)^{\frac{1}{2}}} \approx -\frac{2\pi}{\sqrt{T_0}} \sum_{n=1}^{\infty} a_n \sqrt{n} \sin\left[\frac{2n\pi t}{t_0} - \frac{\pi}{4}\right]$$

$$\frac{p_2(\mathbf{x}, t)}{\rho_0 V^2 \sqrt{M} \cos \Theta (a/|\mathbf{x}|)^{\frac{1}{2}}} \approx \frac{2\pi}{\sqrt{T_0}} \sum_{n=1}^{\infty} b_n \sqrt{n} \cos\left[\frac{2n\pi t}{t_0} - \frac{\pi}{4}\right], \quad |\mathbf{x}| \to \infty,$$

where [ ] denotes evaluation at the retarded time $t - |\mathbf{x}|/c_0$. The corresponding nondimensional pressures are plotted in Fig. 8.1.6 (taking the first 26 terms in the series); both have similar orders of magnitude, and exhibit rapid variations at the retarded times at which the vortex is directly above and below the airfoil.

8.2 Parallel Blade–Vortex Interactions in Three Dimensions

We now examine to what extent the simple two-dimensional methods of the previous section can be adapted to wings of finite span and variable chord for problems of the kind shown in Fig. 8.2.1. The general representation of the sound produced by vortex–airfoil interactions is discussed in Section 7.1, when

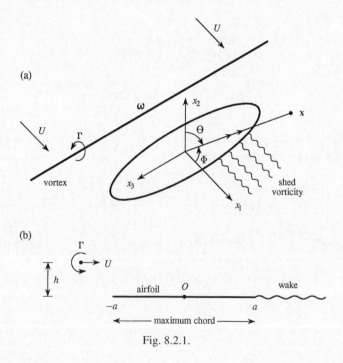

Fig. 8.2.1.

the airfoil chord can be regarded as compact. The general solution is applicable to airfoils of arbitrary span, but we shall consider only the case where the span is compact; predictions for a noncompact span will be intermediate between those discussed here and those in Section 8.1.

Consider a planar airfoil of either rectangular or elliptic planform, orientated as illustrated in Fig. 8.2.1 at zero angle of attack to a mean flow at speed U in the x_1 direction. A spanwise line vortex of strength Γ is swept past the airfoil at an initial standoff distance h above the airfoil, as indicated in the side view of Fig. 8.2.1b. When $h = 0$ it will be necessary to take account of nonlinear interactions with the airfoil.

For an airfoil of compact chord and span the acoustic pressure produced by the interaction is given by Equation (5.4.4):

$$p(\mathbf{x}, t) \approx \frac{-\rho_0 x_j}{4\pi c_0 |\mathbf{x}|^2} \frac{\partial}{\partial t} \int (\boldsymbol{\omega} \wedge \mathbf{v}) \left(\mathbf{y}, t - \frac{|\mathbf{x}|}{c_0}\right) \cdot \nabla Y_j(\mathbf{y}) \, d^3\mathbf{y}, \quad |\mathbf{x}| \to \infty.$$

$$(8.2.1)$$

It is assumed that the section of the line vortex that interacts with the airfoil remains rectilinear, with the representation

$$\boldsymbol{\omega} = \Gamma \mathbf{k} \delta(x_1 - x_{01}(t)) \delta(x_2 - x_{02}(t)), \quad \text{where } \mathbf{x}_0 = (x_{01}, x_{02}, 0).$$

For an elliptic airfoil of span L (between $-\frac{1}{2}L < x_3 < \frac{1}{2}L$), the Kirchhoff vector $\mathbf{Y}$ has the components

$$Y_1 = y_1,$$

$$Y_2 = \begin{cases} \text{Re}(-i\sqrt{z^2 - \hat{a}(y_3)^2}), & |y_3| < \frac{1}{2}L \\ y_2, & |y_3| > \frac{1}{2}L \end{cases}, \quad Y_3 = y_3, \quad z = y_1 + iy_2,$$

where $2\hat{a}(y_3)$ is the airfoil chord at the spanwise location y_3. For the rectangular airfoil $\hat{a}(y_3) \equiv a = \text{constant}$; for the elliptic airfoil $\hat{a}(y_3)$ assumes a maximum value of a at $y_3 = 0$, and we shall write

$$\frac{\hat{a}(y_3)}{a} = \sqrt{1 - \frac{4y_3^2}{L^2}}, \quad |y_3| < \frac{1}{2}L. \tag{8.2.2}$$

Vorticity is shed into the wake of the airfoil in accordance with the Kutta condition of unsteady aerodynamics. This smooths out conditions at the trailing edge, so that sound is generated primarily as the vortex passes over the leading edge of the airfoil. As before, this can be dealt with in a first approximation by ignoring the shed vorticity and deleting the trailing edge singularity of

Green's function, by using the following modification of the x_2 component of **Y**:

$$Y_2 = \text{Re}(\sqrt{2\hat{a}(y_3)}\sqrt{z + \hat{a}(y_3)}), \quad |y_3| < \frac{1}{2}L.$$

Then, (8.2.1) becomes

$$p(\mathbf{x}, t) \equiv p_1(\mathbf{x}, t) + p_2(\mathbf{x}, t)$$

$$\approx \frac{\rho_0 \Gamma \cos \Phi}{4\pi c_0 |\mathbf{x}|} \frac{\partial}{\partial t} \int_{-\frac{L}{2}}^{\frac{L}{2}} \left[\frac{dx_{02}}{dt} \right] dy_3 + \frac{\rho_0 \Gamma \cos \Theta}{4\sqrt{2}\pi c_0 |\mathbf{x}|}$$

$$\times \frac{\partial}{\partial t} \int_{-\frac{L}{2}}^{\frac{L}{2}} \left[\text{Im} \left(\frac{dz_0}{dt} \frac{\sqrt{\hat{a}(y_3)}}{\sqrt{z_0(t) + \hat{a}(y_3)}} \right) \right] dy_3 \quad |\mathbf{x}| \to \infty, \quad (8.2.3)$$

where Φ, Θ are respectively the angles shown in Fig. 8.2.1a between the x_1 and x_2 directions and the radiation direction, $z_0(t) = x_{01}(t) + ix_{02}(t)$, and quantities in square braces are evaluated at the retarded time $[t] = t - |\mathbf{x}|/c_0$.

The first term on the right is the suction force dipole, aligned with the airfoil chord, whose strength is determined by the x_2 component of the vortex convection velocity. It is assumed to be nonzero only over the section $-\frac{1}{2}L < y_3 < \frac{1}{2}L$ of the vortex, where $dx_{02}/dt \neq 0$ because of nonlinear interactions with the airfoil. The second term is the conventional lift dipole radiation. Note that 'infinite' contributions to the integrals from $|y_3| > L/2$ are constant because $\omega \wedge \mathbf{v}$ is constant for $|y_3| > L/2$, and have been discarded (c.f., Section 7.3).

8.2.1 Linear Theory

When there is no back-reaction of the airfoil on the vortex the convection velocity of the vortex is equal to the mean stream velocity

$$\frac{dx_{01}}{dt} = U, \quad \frac{dx_{02}}{dt} = 0.$$

The radiation is produced entirely by the lift dipole, and if the vortex crosses the midchord of the airfoil at time $t = 0$ (8.2.3) reduces to

$$\frac{p_2(\mathbf{x}, t)}{\rho_0 \Gamma U M \cos \Theta(L/|\mathbf{x}|)/4\pi a} \approx -\frac{1}{2^{\frac{3}{2}}} \int_{-\frac{1}{2}}^{\frac{1}{2}} \text{Im} \left(\frac{\sqrt{\frac{\hat{a}}{a}}}{\left(\frac{U[t]}{a} + \frac{\hat{a}}{a} + i\frac{h}{a} \right)^{\frac{3}{2}}} \right) d\hat{y}_3,$$

$$(8.2.4)$$

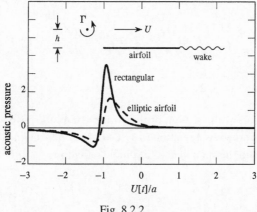

Fig. 8.2.2.

where $\hat{y}_3 = y_3/L$; $\hat{a}/a = 1$ for the rectangular airfoil, and is given by (8.2.2) for the elliptic airfoil. For a rectangular airfoil the integral evaluates to

$$\text{Im}\left\{\left(\frac{U[t]}{a} + 1 + i\frac{h}{a}\right)^{-\frac{3}{2}}\right\}.$$

The acoustic pressure signatures (the left-hand side of (8.2.4)) for the rectangular and elliptic airfoils are plotted in Fig. 8.2.2 for a vortex standoff distance $h = 0.2a$. The profiles are qualitatively similar to the corresponding plot in Fig. 8.1.3 for an airfoil of infinite span, although in three dimensions the amplitude decreases much more rapidly with increasing retarded distance of the vortex from the leading edge. The peak amplitude is larger and the acoustic pulse is of smaller duration $\sim h/U$ for the rectangular airfoil because different sections of the vortex interact with the leading edge of the elliptic airfoil at different times during a total interaction time $\sim a/U > h/U$.

8.2.2 Nonlinear Theory

When the standoff distance $h = 0$, image vorticity in the airfoil prevents the vortex from impinging on the leading edge, and causes the trajectory to be locally deflected above the airfoil (for $\Gamma > 0$). If the leading edge of the airfoil is suitably rounded (so that no additional vortex shedding occurs) this case can be treated for a rectangular airfoil by assuming that only the section of the vortex within the span $-\frac{1}{2}L < x_3 < \frac{1}{2}L$ of the airfoil is affected in this way, and that the distorted path can be approximated by that for locally two-dimensional flow.

Let

$$\epsilon = \frac{\Gamma}{4\pi a U}, \quad T = \frac{Ut}{a}, \quad Z = \frac{z_0}{a}, \quad \frac{dZ}{dT} = u(T) + iv(T),$$

$$\mathcal{W}(Z) = \frac{1}{\sqrt{2}\sqrt{Z+1}}.$$

Then, the motion of the section of the vortex within the airfoil span $(-\frac{1}{2}L < x_3 < \frac{1}{2}L)$ is governed by Equation (8.1.10) (where ζ_0 is defined in terms of z_0 as in (8.1.1)), and the suction and lift dipole radiation pressures are given by

$$\left.\begin{array}{l} \dfrac{p_1(\mathbf{x}, t)}{\rho_0 \Gamma U M \cos \Phi (L/|\mathbf{x}|)/4\pi a} \approx \left[\dfrac{dv}{dT}\right] \\[4mm] \dfrac{p_2(\mathbf{x}, t)}{\rho_0 \Gamma U M \cos \Theta (L/|\mathbf{x}|)/4\pi a} \approx \dfrac{\partial}{\partial T}[\mathrm{Im}\,(\mathcal{W}(Z)(u+iv))] \end{array}\right\} \quad |\mathbf{x}| \to \infty,$$

where [ ] denotes evaluation at the retarded time $t - |\mathbf{x}|/c_0$.

These nondimensional pressures are plotted in Fig. 8.2.3 for a velocity ratio $\epsilon = 0.2$ when the vortex is released upstream with $h = 0$. The upper part of the

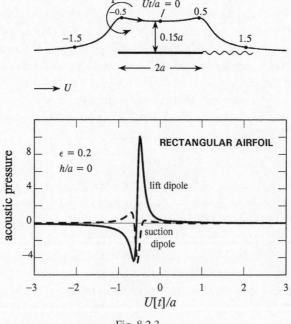

Fig. 8.2.3.

figure shows the path followed by those sections of the vortex inboard of the airfoil tips; it is the same as that depicted in Fig. 8.1.5 for the infinite span airfoil. As in that case, small acoustic blips spuriously predicted during the passage of the vortex past the trailing edge have been removed (c.f., Fig. 8.1.4). The three-dimensional acoustic pulses are narrower than those predicted in two dimensions.

8.3 Vortex Passing over a Spoiler

The sound produced when vorticity interacts at low Mach number with surface irregularities on a nominally plane, rigid wall is produced by dipoles orientated in the plane of the wall, that is, by the unsteady *wall drag*. A simple canonical interaction (Kasoev, 1976) involving a line vortex near a thin vertical spoiler is illustrated in Fig. 8.3.1. The wall coincides with the plane $x_2 = 0$, and the spoiler extends along the x_2 axis from $x_2 = 0$ to $x_2 = a > 0$ for $-\infty < x_3 < \infty$. The vortex

$$\omega = \Gamma \mathbf{k}\delta(\mathbf{x} - \mathbf{x}_0(t)), \quad \text{where } \mathbf{x}_0 = (x_{01}, x_{02}, 0),$$

is parallel to the spoiler, and is assumed to convect over it in a low Mach number, irrotational mean stream having uniform speed U in the x_1 direction.

Define $z = x_1 + ix_2, z_0 = x_{01} + ix_{02}$. The transformation

$$\zeta = \sqrt{\frac{z^2}{a^2} + 1} \tag{8.3.1}$$

maps the fluid region onto the upper half Im $\zeta > 0$ of the ζ-plane. The left and right faces of the spoiler ($x_1 = \mp 0$) transform respectively into the intervals $-1 < \zeta < 0$ and $0 < \zeta < 1$ of the real ζ axis, and the vortex maps into an equal vortex at $\zeta = \zeta_0$. The mean flow is parallel to the real axis in the ζ plane,

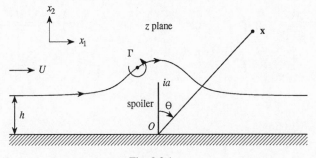

Fig. 8.3.1.

with complex potential $Ua\zeta$. The complex potential of the whole flow in the ζ plane is obtained by introducing an image vortex of strength $-\Gamma$ at the complex conjugate point $\zeta = \zeta_0^*$, and is given by

$$w(\zeta) = -\frac{i\Gamma}{2\pi} \ln(\zeta - \zeta_0) + \frac{i\Gamma}{2\pi} \ln(\zeta - \zeta_0^*) + Ua\zeta.$$

Hence, setting $\zeta = \zeta(z)$ the motion in the z plane is defined by

$$w(z) = -\frac{i\Gamma}{2\pi} \ln(\zeta(z) - \zeta(z_0)) + F(z),$$

$$\text{where } F(z) = \frac{i\Gamma}{2\pi} \ln(\zeta(z) - \zeta(z_0)^*) + Ua\zeta(z),$$

which is in the form (4.6.1). The equation of motion of the vortex at z_0 is therefore (see (4.6.3))

$$\frac{dz_0^*}{dt} \equiv \frac{dx_{01}}{dt} - i\frac{dx_{02}}{dt} = -\frac{i\Gamma\zeta''(z_0)}{4\pi\zeta'(z_0)} + F'(z_0),$$

that is,

$$\frac{dZ^*}{dT} = -i\left\{ \frac{1}{Z(Z^2 + 1)} - \frac{2Z}{Z^2 + 1 - |Z^2 + 1|} \right\} + \frac{\epsilon Z}{\sqrt{Z^2 + 1}}, \quad (8.3.2)$$

$$\text{where } \quad Z = \frac{z_0}{a}, \qquad T = \frac{Vt}{a}, \qquad V = \frac{\Gamma}{4\pi a}, \qquad \epsilon = \frac{U}{V}.$$

The compact Green's function for this problem (applicable when the acoustic wavelength $\gg \eth$) is given by (6.1.6), so that the analogue of Equation (6.2.4) for the far-field acoustic pressure becomes

$$p(\mathbf{x}, t)$$

$$\approx \frac{-\rho_0\Gamma x_1}{\pi\sqrt{2c_0}|\mathbf{x}|^{\frac{3}{2}}} \frac{\partial}{\partial t} \int_{-\infty}^{t-|\mathbf{x}|/c_0} \left(\mathbf{k} \wedge \frac{d\mathbf{x}_0}{d\tau}(\tau) \cdot \nabla Y_1(\mathbf{x}_0(\tau)) \right) \frac{d\tau}{\sqrt{t - \tau - |\mathbf{x}|/c_0}}$$

$$= \frac{-\rho_0\Gamma x_1}{\pi\sqrt{2c_0}|\mathbf{x}|^{\frac{3}{2}}} \frac{\partial}{\partial t} \int_{-\infty}^{t-|\mathbf{x}|/c_0} \left\{ \frac{dx_{01}}{d\tau}\frac{\partial Y_1}{\partial y_2} - \frac{dx_{02}}{d\tau}\frac{\partial Y_1}{\partial y_1} \right\}_{\mathbf{x}_0(\tau)} \frac{d\tau}{\sqrt{t - \tau - |\mathbf{x}|/c_0}},$$

$$(8.3.3)$$

where the Kirchhoff vector

$$Y_1 = \operatorname{Re}(a\zeta) = a\operatorname{Re}(\sqrt{Z^2 + 1}) \quad \text{at} \quad z = z_0. \quad (8.3.4)$$

The radiation is produced by the unsteady drag force F_1 exerted on the fluid by the spoiler, given (per unit span) by

$$F_1 = -\rho_0 \int \boldsymbol{\omega} \wedge \mathbf{v} \cdot \nabla Y_1 \, dy_1 \, dy_2 = -\rho_0 \Gamma \mathbf{k} \wedge \frac{d\mathbf{x}_0}{dt} \cdot \nabla Y_1(\mathbf{x}_0).$$

This force vanishes, and therefore no sound is generated, in the linearized approximation in which the vortex is assumed to translate at the local mean stream velocity, because in that case

$$\frac{d\mathbf{x}_0}{dt} = U \nabla Y_1(\mathbf{x}_0) \quad \text{and} \quad \mathbf{k} \wedge \nabla Y_1 \cdot \nabla Y_1 \equiv 0.$$

Following the procedure of Section 8.1, introduce the notations

$$\frac{dZ}{dT} = u(T) + iv(T), \qquad \mathcal{W} = \frac{d}{dz}(\sqrt{z^2 + a^2}) = \frac{Z}{\sqrt{Z^2 + 1}} \qquad (8.3.5)$$

evaluated at the vortex, and make the substitution $\hat{T} = V\tau/a$ in (8.3.3) to obtain the acoustic pressure in the form

$$p(\mathbf{x}, t) \approx \frac{\rho_0 \Gamma V \sqrt{M} \sin \Theta}{a\pi \sqrt{2}} \left(\frac{a}{|\mathbf{x}|}\right)^{\frac{1}{2}} \frac{\partial}{\partial T} \int_{-\infty}^{[T]} \mathrm{Im}\left(\mathcal{W}(Z)\frac{dZ}{d\hat{T}}\right) \frac{d\hat{T}}{\sqrt{[T] - \hat{T}}},$$

that is,

$$\frac{p(\mathbf{x}, t)}{\rho_0 V^2 \sqrt{M} \sin \Theta (a/|\mathbf{x}|)^{\frac{1}{2}}} \approx 2^{\frac{5}{2}} \frac{\partial}{\partial T} \int_0^{\infty} \mathrm{Im}(\mathcal{W}(Z)(u + iv))([T] - \lambda^2) \, d\lambda,$$

$$(8.3.6)$$

where $[T] = V[t]/a$ is the nondimensional retarded time and $M = V/c_0$.

The vortex path equation (8.3.2) and the acoustic pressure integral (8.3.6) are evaluated numerically, taking the initial position of the vortex to be several spoiler heights a upstream, where its motion is unaffected by the spoiler. The upper part of Fig. 8.3.2 shows the vortex trajectories when the initial distance of the vortex from the wall is $h = 0.75a$ for the two cases (i) of no mean flow, $U = 0$, and (ii) $U = V \equiv \Gamma/4\pi a$; the corresponding nondimensional acoustic pressures (8.3.6) are plotted in the lower part of the figure, where time is measured from the instant that the vortex passes the spoiler. The effect of mean flow is to draw the trajectory marginally closer to the spoiler as it passes the tip of the spoiler where the interaction is strongest. The convection velocity at this point is also increased from about $1.98V$ when $U = 0$ to $3.95V$ when $U = V$, and this is responsible for more than doubling the amplitude and the effective frequency of the sound.

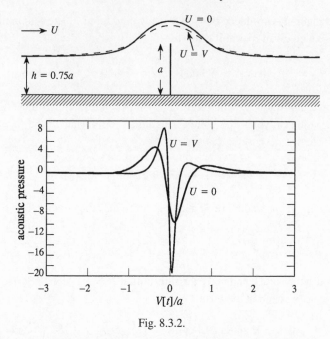

Fig. 8.3.2.

8.4 Bluff Body Interactions: The Circular Cylinder

Low Mach number, two-dimensional interactions of a line vortex with a circular cylinder provide an interesting contrast to the sharp-edge problems discussed above. Let the cylinder have radius a and be coaxial with the x_3 axis, and let there be an irrotational mean flow at speed U past the cylinder in the x_1 direction, with no mean circulation about the cylinder.

Set $z = x_1 + ix_2$ and let the vortex of strength Γ have the complex position $z_0 = x_{01} + ix_{02}$ at time t. The complex potential $w(z)$ is found by placing an image vortex $-\Gamma$ at the *inverse point* $z = a/z_0^*$ within the cylinder, a vortex $+\Gamma$ at the centre to make the circulation vanish, and by adding the potential for the uniform mean flow past the cylinder (c.f., Section 8.1):

$$w(z) = -\frac{i\Gamma}{2\pi}\ln(z-z_0) + \frac{i\Gamma}{2\pi}\ln\left(z - \frac{a^2}{z_0^*}\right) - \frac{i\Gamma}{2\pi}\ln z + U\left(z + \frac{a^2}{z}\right). \quad (8.4.1)$$

The velocity potential governing the motion of the vortex at z_0 is obtained by deleting the self-potential

$$-\frac{i\Gamma}{2\pi}\ln(z - z_0).$$

Hence we arrive at the equation of motion

$$\frac{dZ^*}{dT} = \frac{i}{Z(|Z|^2 - 1)} + \epsilon\left(1 - \frac{1}{Z^2}\right),$$ (8.4.2)

$$\text{where} \quad Z = \frac{z_0}{a}, \quad V = \frac{\Gamma}{2\pi a}, \quad T = \frac{Vt}{a}, \quad \epsilon = \frac{U}{V},$$

$$\text{and} \quad \frac{dz_0}{dt} = V\frac{dZ}{dT} \equiv V(u + iv).$$

8.4.1 The Acoustic Pressure

The far-field sound produced by the vortex is calculated from (6.2.4):

$$p(\mathbf{x}, t) \approx \frac{-\rho_0\Gamma x_j}{2\pi\sqrt{2c_0}|\mathbf{x}|^{\frac{3}{2}}} \frac{\partial}{\partial t} \int_{-\infty}^{t-|\mathbf{x}|/c_0} \left\{\frac{dx_{01}}{d\tau}\frac{\partial Y_j}{\partial y_2} - \frac{dx_{02}}{d\tau}\frac{\partial Y_j}{\partial y_1}\right\}_{\mathbf{x}_0(\tau)}$$

$$\times \frac{d\tau}{\sqrt{t - \tau - |\mathbf{x}|/c_0}},$$ (8.4.3)

where the components of the Kirchhoff vector can be written (see Section 4.5)

$$Y_1 = \text{Re}\left(z + \frac{a^2}{z}\right), \quad Y_2 = \text{Re}\left\{-i\left(z - \frac{a^2}{z}\right)\right\}, \quad z = y_1 + iy_2. \quad (8.4.4)$$

By defining

$$\mathcal{W}_1 = \frac{d}{dz}\left(z + \frac{a^2}{z}\right) \equiv 1 - \frac{1}{Z^2}, \quad \mathcal{W}_2 = \frac{d}{dz}\left\{-i\left(z - \frac{a^2}{z}\right)\right\} \equiv -i\left(1 + \frac{1}{Z^2}\right)$$

evaluated at z_0, and making the change of integration variable $\hat{T} = V\tau/a$, Equation (8.4.3) can be written

$$p(\mathbf{x}, t) \approx \frac{\rho_0\Gamma V\sqrt{M}x_j}{2\pi\sqrt{2a}|\mathbf{x}|^{\frac{3}{2}}} \frac{\partial}{\partial T} \int_{-\infty}^{[T]} \text{Im}(\mathcal{W}_j(u + iv))(\hat{T})\frac{d\hat{T}}{\sqrt{[T] - \hat{T}}},$$

$$\text{where} \quad M = \frac{V}{c_0}, \quad [T] = \frac{V}{a}\left(t - \frac{|\mathbf{x}|}{c_0}\right).$$

The subscripts $j = 1, 2$ in this formula respectively correspond to the acoustic pressures p_1, p_2, say, produced by *drag* and *lift* dipoles, whose strengths are determined by the force (F_1, F_2) exerted on the fluid (per unit span) by the cylinder. The integrals must be evaluated numerically using the numerical

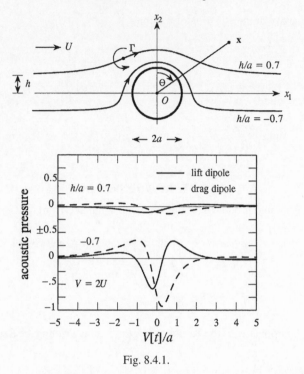

Fig. 8.4.1.

solution of Equation (8.4.2) for the vortex path. This is done by making the further change of integration variable $\lambda = \sqrt{[T] - \hat{T}}$, in which case

$$\frac{p_1(\mathbf{x}, t)}{\rho_0 V^2 \sqrt{M} \sin \Theta (a/|\mathbf{x}|)^{\frac{1}{2}}} \approx 2^{\frac{1}{2}} \frac{\partial}{\partial T} \int_0^\infty \mathrm{Im}(\mathcal{W}_1(u + iv))([T] - \lambda^2) \, d\lambda,$$

$$\frac{p_2(\mathbf{x}, t)}{\rho_0 V^2 \sqrt{M} \cos \Theta (a/|\mathbf{x}|)^{\frac{1}{2}}} \approx 2^{\frac{1}{2}} \frac{\partial}{\partial T} \int_0^\infty \mathrm{Im}(\mathcal{W}_2(u + iv))([T] - \lambda^2) \, d\lambda.$$

The calculation begins at time T', say, by taking the initial position of the vortex to be far upstream of the cylinder at $z_0 = -L + ih$, where $L \gg a$ is sufficiently large that the source strengths are negligible for $T < T'$ (Fig. 8.4.1). The upper limits of integration are then finite because the source terms vanish as soon as $[T] - \lambda^2 < T'$.

Figure 8.4.1 illustrates the typical nondimensional waveforms produced when $V \equiv \Gamma/2\pi a = 2U$ and for $h/a = \pm 0.7$, time being measured from the instant that the vortex crosses $x_1 = 0$. The amplitude of the sound decreases rapidly with increasing distance of closest approach of the vortex to the cylinder; near the cylinder the translational velocity of the vortex is increased because the mean flow velocity is larger, and also because of the increased influence of the image vorticity. In cases where $U \gg V$, the lift dipole will tend to predominate because convection by the image vortices can then be neglected in a first approximation, and the drag $\sim (\omega \wedge U \nabla Y_1) \cdot \nabla Y_1 \equiv 0$.

8.4.2 Wall-Mounted Cylinder

The case of ideal motion of a vortex translating past a cylindrical, semicircular projection on a rigid wall (Fig. 8.4.2) can be treated by the method used for the spoiler in Section 8.3. The problem is equivalent to that in which a *vortex pair*, consisting of a vortex Γ at z_0 accompanied by an image of strength $-\Gamma$ at z_0^*, is incident symmetrically on a circular cylinder. In this case, the lift dipole vanishes identically.

The velocity potential of the unsteady motion is given by augmenting the complex potential (8.4.1) by the terms

$$\frac{i\Gamma}{2\pi} \ln(z - z_0^*) - \frac{i\Gamma}{2\pi} \ln\left(z - \frac{a^2}{z_0}\right) + \frac{i\Gamma}{2\pi} \ln z,$$

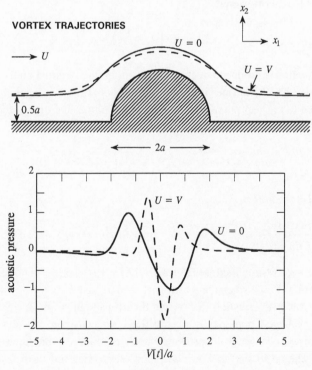

Fig. 8.4.2.

which correspond to the net potential produced by the image. Then, the equation of motion becomes

$$\frac{dZ^*}{dT} = i\left\{\frac{1}{Z-Z^*} + \frac{Z-Z^*}{(|Z|^2-1)(Z^2-1)}\right\} + \epsilon\left(1-\frac{1}{Z^2}\right), \quad (8.4.5)$$

where $Z = \dfrac{z_0}{a}, \quad V = \dfrac{\Gamma}{2\pi a}, \quad T = \dfrac{Vt}{a}, \quad \epsilon = \dfrac{U}{V}.$

The compact Green's function is given by (6.1.6), and the far-field acoustic pressure by

$p(\mathbf{x}, t)$

$$\approx \frac{-\rho_0\Gamma x_1}{\pi\sqrt{2c_0}|\mathbf{x}|^{\frac{3}{2}}} \frac{\partial}{\partial t} \int_{-\infty}^{t-|\mathbf{x}|/c_0} \left(\mathbf{k} \wedge \frac{d\mathbf{x}_0}{d\tau}(\tau) \cdot \nabla Y_1(\mathbf{x}_0(\tau))\right) \frac{d\tau}{\sqrt{t-\tau-|\mathbf{x}|/c_0}}$$

$$= \frac{-\rho_0\Gamma x_1}{\pi\sqrt{2c_0}|\mathbf{x}|^{\frac{3}{2}}} \frac{\partial}{\partial t} \int_{-\infty}^{t-|\mathbf{x}|/c_0} \left\{\frac{dx_{01}}{d\tau}\frac{\partial Y_1}{\partial y_2} - \frac{dx_{02}}{d\tau}\frac{\partial Y_1}{\partial y_1}\right\}_{\mathbf{x}_0(\tau)} \frac{d\tau}{\sqrt{t-\tau-|\mathbf{x}|/c_0}},$$
$$(8.4.6)$$

where the Kirchhoff vector

$$Y_1 = \text{Re}\left(z + \frac{a^2}{z}\right). \quad (8.4.7)$$

The radiation is produced by the unsteady drag force exerted on the fluid by the cylinder, which vanishes in the linearized approximation, when the vortex is assumed to convect passively at the local velocity of the undisturbed mean stream. As before, set

$$\frac{dZ}{dT} = u(T) + iv(T), \quad \mathcal{W}_1 = \frac{d}{dz}\left(z + \frac{a^2}{z}\right) = 1 - \frac{1}{Z^2} \quad (8.4.8)$$

evaluated at the vortex. Then,

$$\frac{p(\mathbf{x}, t)}{\rho_0 V^2\sqrt{M}\sin\Theta(a/|\mathbf{x}|)^{\frac{1}{2}}} \approx 2^{\frac{3}{2}}\frac{\partial}{\partial T}\int_0^\infty \text{Im}(\mathcal{W}_1(u+iv))([T]-\lambda^2)\,d\lambda, \quad (8.4.9)$$

where the angle Θ is defined as in Fig. 8.4.1, $[T] = V[t]/a$ is the nondimensional retarded time ($T = 0$ when the vortex is at $x_1 = 0$), and $M = V/c_0$.

The vortex path equation (8.4.5) and the acoustic pressure integral (8.4.9) must be evaluated numerically, taking the initial position of the vortex to be several cylinder radii a upstream where its motion is unaffected by the cylinder. The upper part of Fig. 8.4.2 shows the vortex trajectories when the initial standoff distance of the vortex from the wall $h = 0.5a$ for the two cases (i) of no mean flow, $U = 0$, and (ii) $U = V$. The vortex convection velocity at $y_1 = 0$ is increased from about $1.23V$ when $U = 0$ to $3.07V$ when $U = V$; this is responsible for the increased acoustic amplitude and for more than doubling the effective frequency. The waveforms and these general conclusions are qualitatively similar to those discussed in Section 8.3 for the sharp-edged spoiler.

8.5 Vortex Ring and Sphere

Perhaps the simplest low Mach number, inviscid, three-dimensional vortex–surface interaction amenable to analysis is the axisymmetric motion of a ring vortex over a sphere. Let a sphere of radius a be placed with its centre at the origin in the presence of a uniform mean flow at speed U in the x_1 direction. A vortex ring of radius $r_0(t)$ and circulation Γ is coaxial with the x_1 axis and translates in the positive x_1 direction under the influence of the mean flow, self-induction and image vorticity in the sphere (Fig. 8.5.1). We shall assume the vortex core is circular (and remains circular throughout the interaction) with radius $\sigma(t) \ll r_0$. The self-induced velocity of the ring (in inviscid flow) is parallel to the x_1-axis at speed u_Γ given approximately by Kelvin's formula (Batchelor, 1967)

$$u_\Gamma(r_0, \sigma) = \frac{\Gamma}{4\pi r_0} \left\{ \ln\left(\frac{8r_0}{\sigma}\right) - \frac{1}{4} \right\}. \qquad (8.5.1)$$

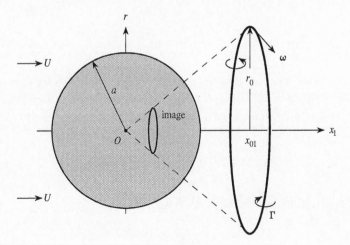

Fig. 8.5.1.

The image vorticity consists of a coaxial ring vortex whose circulation Γ', radius r_0', and axial location x_{01}' are given by

$$\Gamma' = -\frac{\Gamma\left(r_0^2 + x_{01}^2\right)^{\frac{1}{2}}}{a}, \qquad r_0' = \frac{a^2 r_0}{r_0^2 + x_{01}^2}, \qquad x_{01}' = \frac{a^2 x_{01}}{r_0^2 + x_{01}^2}, \qquad (8.5.2)$$

where the planes of symmetry of the ring vortex and its image cut the x_1 axis respectively at $x_{01}(t)$, $x_{01}'(t)$. The motion of the vortex produced by the combined induction by the image and the mean flow can be expressed in terms of the *Stokes stream function* $\psi(\dot{r}, x_1)$ (Batchelor, 1967; Ting and Klein, 1991)

$$\psi(r, x_1) = \frac{U r^2}{2}\left(1 - \frac{a^3}{\left(r^2 + x_1^2\right)^{\frac{3}{2}}}\right) + \frac{\Gamma'}{2\pi}(\mathfrak{R}_+ + \mathfrak{R}_-)\{K(\Lambda) - E(\Lambda)\},$$

$$(8.5.3)$$

where $\mathfrak{R}_\pm = \sqrt{(r \mp r_0')^2 + (x_1 - x_{01}')^2}$, $\qquad \Lambda = \dfrac{\mathfrak{R}_- - \mathfrak{R}_+}{\mathfrak{R}_- + \mathfrak{R}_+}$,

$$K(\Lambda) = \int_0^{\frac{\pi}{2}} \frac{d\mu}{\sqrt{1 - \Lambda^2 \sin^2 \mu}}, \qquad E(\Lambda) = \int_0^{\frac{\pi}{2}} \sqrt{1 - \Lambda^2 \sin^2 \mu}\, d\mu,$$

where r denotes perpendicular distance from the x_1 axis, and $K(\Lambda)$, $E(\Lambda)$ are respectively complete elliptic integrals of the first and second kinds.

The radius $r_0(t)$ and axial position $x_{01}(t)$ of the ring vortex are then determined by the equations of motion

$$\frac{dr_0}{dt} = -\frac{1}{r_0}\frac{\partial \psi}{\partial x_1}(r_0, x_{01}), \qquad \frac{dx_{01}}{dt} = u_\Gamma(r_0, \sigma) + \frac{1}{r_0}\frac{\partial \psi}{\partial r}(r_0, x_{01}). \quad (8.5.4)$$

The core radius σ decreases when r_0 increases, because the vortex lines move with the fluid particles. If $r_0 = h$ and $\sigma = \sigma_0$ are the initial values when the vortex ring is far from the sphere, then at any time t

$$(2\pi r_0)\pi\sigma^2 = (2\pi h)\pi\sigma_0^2, \quad \text{i.e.,} \quad \sigma(t) = \sigma_0\sqrt{\frac{h}{r_0(t)}}$$

so that the self-induced velocity (8.5.1) becomes

$$u_\Gamma = \frac{\Gamma}{4\pi r_0}\left\{\ln\left(\frac{8h}{\sigma_0}\left[\frac{r_0(t)}{h}\right]^{\frac{3}{2}}\right) - \frac{1}{4}\right\}. \qquad (8.5.5)$$

The equations of motion of the vortex are cast in nondimensional terms by defining

$$X = \frac{x_{01}}{a}, \qquad R = \frac{r_0}{a}, \qquad V = \frac{\Gamma}{2\pi a}, \qquad T = \frac{Vt}{a}, \qquad \epsilon = \frac{U}{V}.$$

Then,

$$\frac{dR}{dT} = -\frac{1}{R}\frac{\partial \Psi}{\partial X}(R, X), \quad \frac{dX}{dT} = \frac{1}{R}\frac{\partial \Psi}{\partial R}(R, X) + \frac{1}{2R}\left\{\ln\left(\frac{8h}{\sigma_0}\left[\frac{aR}{h}\right]^{\frac{3}{2}}\right) - \frac{1}{4}\right\},$$

(8.5.6)

$$\text{where} \quad \Psi = \frac{\epsilon R^2}{2}\left(1 - \frac{1}{(R^2 + X^2)^{\frac{3}{2}}}\right) - (R^2 + X^2)^{\frac{1}{2}}$$

$$\times (\hat{\Re}_+ + \hat{\Re}_-)\{K(\Lambda) - E(\Lambda)\},$$

$$\hat{\Re}_\pm = \sqrt{(R \mp R')^2 + (X - X')^2}, \qquad \Lambda = \frac{\hat{\Re}_- - \hat{\Re}_+}{\hat{\Re}_- + \hat{\Re}_+},$$

$$R' = \frac{R}{R^2 + X^2}, \qquad X' = \frac{X}{R^2 + X^2}.$$

Figure 8.5.2 illustrates the sections in the vertical plane of symmetry of the sphere of two typical vortex trajectories predicted by Equations (8.5.6). In both cases the integration is started five sphere diameters upstream with the following initial values for the vortex ring radius and core radius,

$$h = 0.8a, \qquad \sigma_0 = 0.05h. \tag{8.5.7}$$

The solid and broken-line curves in the figure correspond respectively to $\epsilon = 0$

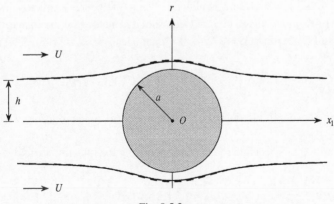

Fig. 8.5.2.

(no mean flow) and $\epsilon = 3$. The latter value is chosen to make the mean stream velocity U approximately the same as the self-induced velocity u_Γ at large distances from the sphere. This has a relatively small effect on the trajectory, although the convection speed of the ring past the sphere is greatly increased.

8.5.1 Acoustic Pressure

When the sphere is acoustically compact, equation (5.4.4) gives

$$p(\mathbf{x}, t) \approx \frac{-\rho_0 x_j}{4\pi c_0 |\mathbf{x}|^2} \frac{\partial}{\partial t} \int (\boldsymbol{\omega} \wedge \mathbf{v}) \left(\mathbf{y}, t - \frac{|\mathbf{x}|}{c_0} \right) \cdot \nabla Y_j(\mathbf{y}) \, d^3\mathbf{y}, \quad |\mathbf{x}| \to \infty. \tag{8.5.8}$$

For the purpose of evaluating the integral we may neglect the finite core size of the vortex, and set

$$\boldsymbol{\omega} = \Gamma \hat{\boldsymbol{\theta}} \delta(r - r_0(t)) \delta(x_1 - x_{01}(t)), \tag{8.5.9}$$

where $\hat{\boldsymbol{\theta}}$ is a unit azimuthal vector, locally tangential to the vorticity $\boldsymbol{\omega}$ and orientated in the *clockwise* direction when the vortex ring is viewed from upstream, as indicated in Fig. 8.5.1.

The force on the sphere is in the mean flow direction – the effective acoustic source is the unsteady *drag* – and only the component

$$Y_1 = y_1 \left(1 + \frac{a^3}{2|\mathbf{y}|^3} \right) \equiv y_1 \left(1 + \frac{a^3}{2 \left(r^2 + y_1^2 \right)^{\frac{3}{2}}} \right) \qquad \text{(from Table 3.9.1)}$$

of the Kirchhoff vector makes a nontrivial contribution to (8.5.8). The production of sound is therefore a nonlinear event – the source explicitly involves only the self-induced velocity and the velocity induced by the image vortex; the mean flow velocity $\mathbf{U} = U \nabla Y_1$ is absent because

$$(\boldsymbol{\omega} \wedge U \nabla Y_1) \left(\mathbf{y}, t - \frac{|\mathbf{x}|}{c_0} \right) \cdot \nabla Y_1(\mathbf{y}) \equiv 0.$$

However, the amplitude and characteristic frequency of the sound both increase with U.

Substituting (8.5.9) into (8.5.8) and evaluating the integral, we find

$$p(\mathbf{x}, t) \approx \frac{-\rho_0 \Gamma \cos \Theta}{2c_0 |\mathbf{x}|} \frac{\partial}{\partial t} \left[r_0 \left(\frac{dx_{01}}{dt} \frac{\partial Y_1}{\partial r} - \frac{dr_0}{dt} \frac{\partial Y_1}{\partial y_1} \right)_{(r_0, x_{01})} \right], \quad |\mathbf{x}| \to \infty,$$

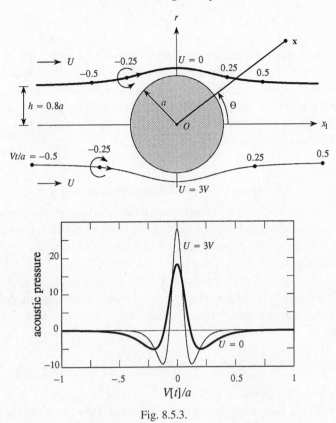

Fig. 8.5.3.

where the quantity in the square braces is evaluated at the retarded position of the vortex ring, and Θ is the angle between the radiation direction and the x_1 axis illustrated in the upper part of Fig. 8.5.3. Expressing this result in nondimensional form, we have

$$
\frac{p(\mathbf{x}, t)}{\rho_0 V^2 M \cos \Theta (a/|\mathbf{x}|)} \approx \pi \frac{\partial}{\partial T} \left[\frac{3 R^2 X}{2(R^2 + X^2)^{\frac{5}{2}}} \frac{dX}{dT} \right.
$$
$$
\left. + \left(1 + \frac{R^2 - 2X^2}{2(R^2 + X^2)^{\frac{5}{2}}} \right) R \frac{dR}{dT} \right],
$$

where $M = V/c_0$, and R and X are the solutions of the vortex equations of motion (8.5.6).

The nondimensional acoustic pressure signatures plotted in Fig. 8.5.3 are for the same the initial conditions (8.5.7) considered above for the vortex ring trajectories in Fig. 8.5.2. The thick solid curve is the pressure profile in the absence of mean flow ($U = 0$). The positions of the vortex ring in this case at several different retarded times $V[t]/a$ are marked on the thick curve in the upper part of Fig. 8.5.3 (time being measured from the instant that the ring crosses the centre of the sphere). Similarly, the thin-line curves in the figure give the pressure signature and retarded positions for $U = 3V$, when the self-induction velocity $u_\Gamma \approx U$ at large distances from sphere. Both the amplitude and frequency of the sound are increased because of the increased convection velocity of the vortex past the sphere.

8.6 Vortex Pair Incident on a Wall Aperture

Hydrodynamic motion in the vicinity of an aperture in a large thin wall generally produces an unsteady volume flux through the aperture, which is acoustically equivalent to a monopole source when the aperture is compact. The upper part of Fig. 8.6.1 depicts a simple model of such a source. The rigid wall coincides with the plane $x_1 = 0$, and is pierced by a two-dimensional slit aperture of width $2a$ whose centerline extends along the x_3 axis. A vortex pair aligned with the x_3 axis, consisting of vortices of strengths $\pm\Gamma$ at the respective complex positions

$$z_0 = x_{01} + ix_{02} \quad \text{and} \quad z_0^* = x_{01} - ix_{02}$$

is incident on the aperture from the left ($x_1 = -\infty$).

The motion is evidently symmetric with respect to the x_1 axis, and the transformation

$$\zeta = \frac{z}{\sqrt{z^2 + a^2}} \quad (z = x_1 + ix_2)$$

maps the region Im $z > 0$ cut along the upper section $x_2 > a$ of the wall onto the upper half of the ζ plane. By the usual method, we accordingly obtain the equation of motion of the vortex pair in the form (Karweit, 1975)

$$\frac{dZ^*}{dT} = \frac{3iZ}{Z^2 + 1} + \frac{2i}{(Z^2 + 1)^{\frac{3}{2}}\{Z/\sqrt{Z^2 + 1} - (Z/\sqrt{Z^2 + 1})^*\}}, \quad (8.6.1)$$

where $Z = \dfrac{z_0}{a}$, $T = \dfrac{Vt}{a}$, $V = \dfrac{\Gamma}{4\pi a}$,

and $\dfrac{dz_0}{dt} = V\dfrac{dZ}{dT} \equiv V(u + iv)$.

Let the initial separation of the vortices at $x_1 = -\infty$ be $2h$. To integrate the equation, we can set $z_0 = -L + ih$ at a convenient initial (but arbitrary) time

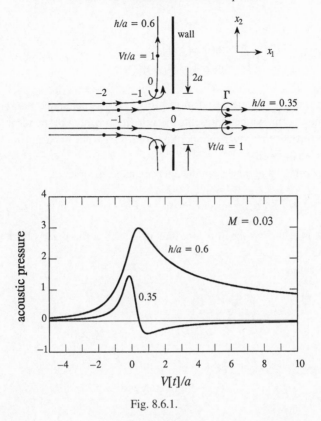

Fig. 8.6.1.

$T = T'$, where $L \gg a$. When h/a is smaller than $2/3^{\frac{3}{2}} \approx 0.385$ the vortex pair passes through the aperture in the manner indicated in Fig. 8.6.1 for $h/a = 0.35$. For larger values of h/a the trajectories of the two vortices separate; the vortices travel along symmetric paths parallel to the wall on either side of the aperture, as illustrated for $h/a = 0.6$.

The acoustic pressure in the far field is given by

$$p(\mathbf{x}, t) \approx -\rho_0 \int \boldsymbol{\omega} \wedge \mathbf{v} \cdot \frac{\partial G}{\partial \mathbf{y}} \, d^2\mathbf{y} \, d\tau, \tag{8.6.2}$$

where G is the compact Green's function (3.9.10) for the wall aperture

$$G(\mathbf{x}, \mathbf{y}, t - \tau) \approx -\frac{\sqrt{c_0} \operatorname{sgn}(x_1)}{\pi \sqrt{2\pi |\mathbf{x}|}} \frac{\chi(t - \tau - |\mathbf{x}|/c_0)}{\sqrt{t - \tau - |\mathbf{x}|/c_0}} \operatorname{Re} \left\{ \ln \left(\frac{\tilde{z}}{a} + \sqrt{\frac{\tilde{z}^2}{a^2} - 1} \right) \right\},$$

$$\tilde{z} = y_2 + iy_1, \tag{8.6.3}$$

and

$$\chi(t) = H(t) \int_0^\infty \frac{\ln(\varpi a\xi^2/4c_0t)e^{-\xi^2}d\xi}{[\ln(\varpi a\xi^2/4c_0t)]^2 + \pi^2}, \qquad \varpi = 1.781072.$$

The dependence on source position **y** in (8.6.3) is contained entirely in the logarithmic term, which represents the velocity potential of the ideal flow that would be produced through the aperture (from left to right) by a uniform pressure drop across the wall.

When vortex shedding from the aperture edges is ignored,

$$\omega = \Gamma\mathbf{k}\delta(y_1 - x_{01})\delta(y_2 - x_{02}) - \Gamma\mathbf{k}\delta(y_1 - x_{01})\delta(y_2 + x_{02}),$$

where **k** is a unit vector in the x_3 direction (out of the plane of the paper in Fig. 8.6.1).

If we define

$$\tilde{\mathcal{W}}(Z) = \left(\frac{1}{\sqrt{\tilde{Z}^2 - 1}}\right)_{\tilde{Z}=iZ^*}, \quad \hat{\chi}(T) = \chi(t), \quad M = \frac{V}{c_0},$$

and put

$$\hat{T} = \frac{V\tau}{a} \text{ in the integral (8.6.2),}$$

we find

$$p(\mathbf{x}, t) \approx \frac{2^{\frac{5}{2}}\rho_0 V^2 \operatorname{sgn}(x_1)}{\sqrt{\pi M}} \left(\frac{a}{|\mathbf{x}|}\right)^{\frac{1}{2}} \int_{-\infty}^{[T]} \operatorname{Re}(\tilde{\mathcal{W}}^*(Z)(u + iv))(\hat{T})$$

$$\times \frac{\hat{\chi}([T] - \hat{T})\,d\hat{T}}{\sqrt{[T] - \hat{T}}}, \quad [T] = \frac{V[t]}{a}.$$

Therefore, by setting $\lambda = \sqrt{[T] - \hat{T}}$ we can write

$$\frac{p(\mathbf{x}, t)}{\rho_0 V^2 \operatorname{sgn}(x_1)(a/|\mathbf{x}|)^{\frac{1}{2}}} \approx \frac{2^{\frac{7}{2}}}{\sqrt{\pi M}} \int_0^\infty \operatorname{Re}(\tilde{\mathcal{W}}^*(Z)(u + iv))([T] - \lambda^2)\hat{\chi}(\lambda^2)\,d\lambda,$$

$$|\mathbf{x}| \to \infty, \quad (8.6.4)$$

where

$$\hat{\chi}(\lambda^2) = \int_0^\infty \frac{\ln(\varpi M\xi^2/4\lambda^2)e^{-\xi^2}d\xi}{[\ln(\varpi M\xi^2/4\lambda^2)]^2 + \pi^2}.$$

As before, the upper limit of integration in (8.6.4) is actually $\lambda = \sqrt{[T] - T'}$, where T' is the nondimensional initial time from which the motion of the vortex pair is calculated.

The value of the integral depends weakly on the characteristic Mach number

$$M = \frac{V}{c_0} \equiv \frac{\Gamma}{4\pi a c_0}.$$

This is just the self-induced convection Mach number of the vortex pair when separated by a distance $2a$. We have taken $M = 0.03$ for the far-field acoustic pressure signatures plotted in Fig. 8.6.1; in air this would imply that $V \sim 10$ m/sec.

The flow induced by the vortex pair approaching the wall forms a localized two-dimensional jet between the vortices, directed toward the wall. The resistance of the wall to this flow causes the pressure just to the left of the wall aperture to rise, forcing fluid through the aperture into the region $x_1 > 0$. The radiation therefore has the characteristics of an acoustic monopole source for $x_1 > 0$ and a sink for $x_1 < 0$. Numerical results are illustrated in the figure for $h/a = 0.35, \ 0.6$. In each case, the time origin has been adjusted to correspond approximately with the peak in the radiated acoustic pressure, which occurs when the vortices pass close to the edges of the aperture. When $h/a = 0.6$ the vortices do not penetrate the aperture but are deflected by the wall; this produces a relatively larger pressure rise than for $h/a = 0.35$, where the vortices pass through the aperture. The maximum acoustic pressure amplitude is found to occur when h/a just exceeds the critical value (~ 0.385), when the vortex trajectories pass very close to the aperture edges. Further increases of h/a beyond 0.6 result in a gradual reduction in the amplitude of the sound, and a corresponding increase in the width of the acoustic pulse (i.e., a decrease in the characteristic frequency of the sound).

Bibliography

Batchelor, G.K. 1967. *An Introduction to Fluid Dynamics.* Cambridge: University Press.

Cannell, P. and Ffowcs Williams, J.E. 1973. Radiation from line vortex filaments exhausting from a two-dimensional semi-infinite duct. *Journal of Fluid Mechanics* **58**: 65–80.

Crighton, D.G. 1972. Radiation from vortex filament motion near a half plane. *Journal of Fluid Mechanics* **51**: 357–362.

Crighton, D.G. 1975. Basic principles of aerodynamic noise generation. *Progress in Aerospace Sciences* **16**: 31–96.

Crighton, D.G. 1985. The Kutta condition in unsteady flow. *Annual Reviews of Fluid Mechanics* **17**: 411–445.

Crighton, D.G., Dowling, A.P., Ffowcs Williams, J.E., Heckl, M., and Leppington, F.G. 1992. *Modern Methods in Analytical Acoustics* (Lecture Notes). London: Springer-Verlag.

Curle, N. 1955. The influence of solid boundaries upon aerodynamic sound. *Proceedings of the Royal Society of London* **A231**: 505–514.

Dhanak, M.R. 1981. Interaction between a vortex filament and an approaching rigid sphere. *Journal of Fluid Mechanics* **110**: 129–147.

Dowling, A.P. and Ffowcs Williams, J.E. 1983. *Sound and Sources of Sound.* Chichester: Ellis Horwood.

Ffowcs Williams, J.E. 1963. The noise from turbulence convected at high speed. *Philosophical Transactions of the Royal Society of London* **A255**: 469–503.

Ffowcs Williams, J.E. and Hawkings, D.L. 1969. Sound generation by turbulence and surfaces in arbitrary motion. *Philosophical Transactions of the Royal Society of London* **A264**: 321–342.

Ffowcs Williams, J.E. and Hall, L.H. 1970. Aerodynamic sound generation by turbulent flow in the vicinity of a scattering half-plane. *Journal of Fluid Mechanics* **40**: 657–670.

Ffowcs Williams, J.E. 1974. Sound production at the edge of a steady flow. *Journal of Fluid Mechanics* **66**: 791–816.

Goldstein, M.E. 1976. *Aeroacoustics.* New York: McGraw-Hill.

Goldstein, S. 1960. *Lectures on Fluid Mechanics.* New York: Interscience.

Howe, M.S. 1975a. Contributions to the theory of aerodynamic sound with application to excess jet noise and the theory of the flute. *Journal of Fluid Mechanics* **71**: 625–673.

Howe, M.S. 1975b. The generation of sound by aerodynamic sources in an inhomogeneous steady flow. *Journal of Fluid Mechanics* **67**: 579–610.

Howe, M.S. 1989. On unsteady surface forces, and sound produced by the normal chopping of a rectilinear vortex. *Journal of Fluid Mechanics* **206**: 131–153.

Howe, M.S. 1998a. *Acoustics of Fluid–Structure Interactions.* Cambridge: Cambridge University Press.

Howe, M.S. 1998b. The compression wave produced by a high-speed train entering a tunnel. *Proceedings of the Royal Society* **A454**: 1523–1534.

Howe, M.S. 2001. Vorticity and the theory of aerodynamic sound. *Journal of Engineering Mathematics* **41**: 367–400.

Howe, M.S., Iida, M., Fukuda, T., and Maeda, T. 2000. Theoretical and experimental investigation of the compression wave generated by a train entering a tunnel with a flared portal. *Journal of Fluid Mechanics* **425**: 111–132.

Karweit, M. 1975. Motion of a vortex pair approaching an opening in a boundary. *Physics of Fluids* **18**: 1604–1606.

Kasoev, S.G. 1976. Sound radiation from a linear vortex over a plane with a projecting edge. *Soviet Physics Acoustics* **22**: 71–72.

Kelvin, Lord. 1867. On vortex motion. *Transactions of the Royal Society of Edinburgh* **25**: 217–260.

Knio, O.M., Ting, L., and Klein, R. 1998. Interaction of a slender vortex with a rigid sphere: Dynamics and far field sound. *Journal of the Acoustical Society of America* **103**: 83–98.

Lamb, Horace. 1932. *Hydrodynamics.* 6th ed. Cambridge: Cambridge University Press.

Landau, L.D. and Lifshitz, E.M. 1987. *Fluid Mechanics.* 2nd ed. Oxford: Pergamon.

Lighthill, M.J. 1952. On sound generated aerodynamically. Part I: General theory. *Proceedings of the Royal Society of London* **A211**: 564–587.

Lighthill, M.J. 1956. The image system of a vortex element in a rigid sphere. *Proceedings of the Cambridge Philosophical Society* **52**: 317–321.

Lighthill, M.J. 1958. *An Introduction to Fourier Analysis and Generalised Functions.* Cambridge: Cambridge University Press.

Lighthill, M.J. 1963. *Laminar Boundary Layers.* Edited by L. Rosenhead. Chs. 1, 2. Oxford: University Press.

Lighthill, James. 1978. *Waves in Fluids.* Cambridge: Cambridge University Press.

Lighthill, J. 1986. *An Informal Introduction to Theoretical Fluid Mechanics.* Oxford: Clarendon Press.

Maeda, T., Matsumura, T., Iida, M., Nakatani, K., and Uchida, K. 1993. Effect of shape of train nose on compression wave generated by train entering tunnel. *Proceedings of the International Conference on Speedup Technology for Railway and Maglev Vehicles.* Japan Society of Mechanical Engineers (Yokohama, Japan 22–26 November) pp. 315–319.

Milne-Thomson, L.M. 1968. *Theoretical Hydrodynamics.* 5th ed. London: Macmillan.

Moore, D.W. 1980. The velocity of a vortex ring with a thin core of elliptical cross-section. *Proceedings of the Royal Society of London* **A370**: 407–415.

Möhring, W. 1978. On vortex sound at low Mach number. *Journal of Fluid Mechanics* **85**: 685–691.

Möhring, W. 1980. Modelling low Mach number noise. *Mechanics of Sound Generation in Flows.* Edited by E.-A. Müller, pp. 85–96. Berlin: Springer-Verlag.

Powell, A. 1960. Aerodynamic noise and the plane boundary. *Journal of the Acoustical Society of America* **32**: 962–990.

Powell, A. 1963. *Mechanisms of Aerodynamic Sound Production*. AGARD Report No. 466.

Powell, A. 1964. Theory of vortex sound. *Journal of the Acoustical Society of America* **36**: 177–195.

Rayleigh, Lord. 1945. *Theory of Sound*. 2 vols. New York: Dover.

Saffman, P.G. 1993. *Vortex Dynamics*. Cambridge: University Press.

Sears, W.R. 1941. Some aspects of non-stationary airfoil theory and its practical applications. *Journal of the Aeronautical Sciences* **8**: 104–108.

Ting, L. and Klein, R. 1991. Viscous vortical flows. *Lecture Notes in Physics*, Vol. 374. New York: Springer-Verlag.

Index